Fundamentals of Molecular Biology and Plant Biotechnology

The Authors

Dr Phundan Singh hails from a farming family of Western Uttar Pradesh. He graduated in Agriculture from Agra University, Agra and obtained his Master and Doctorate degrees from Kanpur University, Kanpur in Genetics and Plant Breeding. Dr. Singh has total experience of more than 40 years in the field of Plant Breeding and Genetics. He has 300 publications to his credit. He has authored 60 books, 18 technical bulletins and contributed 14 chapters in various books. He is an Expert member of Selection Committees and M. Sc. and Ph.D Examiner in different Agricultural Universities. He received lifetime award from CRDA in 2018. He has visited several countries such as Russia, Belarus, Ukraine, Crimea, Uzbekistan, Canada and USA.

Dr Sushma Tiwari obtained her B.Sc. and M.Sc. (Biotechnology) degrees from Awadhesh Pratap Singh University, Rewa, Madhya Pradesh and Ph.D in Life Sciences with specialization in Agricultural Biotechnology from Devi Ahilya University, Indore, [Madhya Pradesh]. She has worked for 10 year as SRF and RA in Plant Biotechnology at IARI, New Delhi. Presently, she is working as Scientist for more than 4 year in the Department of Molecular Biology and Plant Biotechnology, College of Agriculture, Rajmata Vijayeraje Scindia Krishi Vishwa Vidyalaya, Gwalior, Madhya Pradesh. She has more than 35 research papers to her credit published in various reputed journals. She has participated in more than 25 National and International Conferences, Seminars and Workshops. She has received four awards, viz. Emerging Scientist Award, Distinguished Scientist Award, Scientist of the year award and Young scientist award from different scientific societies.

Dr Mridula Billore obtained M.Sc. and Ph.D degrees from JNKVV, Jabalpur in Genetics and Plant Breeding. Dr Billore has more than 30 year experience of teaching, research and extension and served at JNKVV, Jabalpur/ RVSKVV, Gwalior as Assistant Professor, Associate Professor and Dean College of Agriculture. Presently, she is serving as the Dean Faculty at the RVSKVV, Gwalior. She has published more than 70 research papers and 2 book. She received best voluntary pulse research centre award of ICAR in 2010.

Fundamentals of Molecular Biology and Plant Biotechnology

As per ICAR 5th Deans' Committee Recommendation

— *Authors* —

Dr. Phundan Singh

Former Director
ICAR-Central Institute for Cotton Research,
Nagpur – 440 010, Maharashtra

Dr. Sushma Tiwari

Scientist [Biotechnology]
RVSKVV, Gwalior, Madhya Pradesh

Dr. Mridula Billore

Dean Faculty
College of Agriculture
RVSKVV, Gwalior, Madhya Pradesh

2020

Daya Publishing House®

A Division of

Astral International Pvt. Ltd.

New Delhi – 110 002

ISBN : 9789390371204 (Int. Edition)

Published by : **Daya Publishing House®**
A Division of
Astral International Pvt. Ltd.
– ISO 9001 : 2015 Certified Company –
4736/23, Ansari Road, Darya Ganj
New Delhi-110 002
Ph. 011-43549197, 23278134
E-mail: info@astralint.com
Website: www.astralint.com

Dedicated to following Family members of Senior Author

Engineer Rajeev Singh [Son]
Rishabh Singh [Grandson]
Samrat Choudhury [Grandson]
Shaan Choudhury [Grandson]

Preface

Plant biotechnology is an integral part of plant breeding. It has wide practical applications in crop improvement. In India, now Molecular Biology and Plant Biotechnology is offered as a subject both at undergraduate and postgraduate levels in all Conventional and Agricultural Universities. There are several books on the subject both of foreign and Indian origin. However, the course requirement of undergraduate students is not fulfilled by any single source and students have to go through voluminous literature to cover their course requirements. Present book has been designed to fulfil this long felt need of Indian students.

The book has been divided into two sections. Section 1 deals with Molecular Biology in its 8 chapters. Section 2 consists of 14 chapters covering Plant Biotechnology. Each chapter has been presented pointwise and in step by step manner for easy grasping of undergraduate students. Glossary of technical terms is presented at the end of each chapter. Some useful references are given at the end for gathering further details.

The information contained in this book has been gathered from various published sources and Internet websites. Attempts have been made to provide latest information even then some valuable information might have been missed.

The cooperation extended by the wife of senior author Mrs. Jaswanti Singh during preparation of this book is highly appreciable. Hope, this Volume would be useful to the students, researchers, and teachers engaged in the field of molecular biology and plant biotechnology. Constructive suggestions of students and teachers are invited for further improvement of this book.

Dr. Phundan Singh

Dr. Sushma Tiwari

Dr. Mridula Billore

Contents

Section II – Plant Biotechnology

Syllabus
[As per 5th Dean's Committee Report of ICAR]

Nucleic acids: Importance and classification; Structure of Nucleotides, A, B and Z DNA; RNA: Types and Secondary and Tertiary structure.

Concepts and applications of plant biotechnology: Scope, organ culture, embryo culture, cell suspension culture, callus culture, anther-culture, pollen culture and ovule culture and their applications; Micro-propagation methods; organogenesis and embryogenesis, Synthetic seeds and their significance; Embryo rescue and its significance; somatic hybridization and cybrids; Somaclonal variation and its use in crop improvement; cryo-preservation; Introduction to recombinant DNA methods: physical (Gene gun method), chemical (PEG mediated) and Agrobacterium mediated gene transfer methods; Transgenics and its importance in crop improvement; PCR techniques and its applications; RFLP, RAPD, SSR; Marker Assisted Breeding in crop improvement; Biotechnology regulations.

Unitization of Syllabus

1. History of molecular biology and biotechnology;
2. Nucleic acids: Importance and classification; Structure of Nucleotides, A, B and Z DNA;
3. RNA: Types and Secondary and Tertiary structure;
4. Concepts and applications of plant biotechnology;
5. Basics of Gene regulation: The operon concept (Lac operon, Tryp operon)
6. Principal of plant tissue culture, Scope;
7. Organ culture, embryo culture, cell suspension culture, callus culture, anther culture, pollen culture and ovule culture and their applications
8. Micro-propagation methods; organogenesis and embryogenesis, Synthetic seeds and their significance;

9. Embryo rescue and its significance;
10. Somatic hybridization and cybrids; Somaclonal variation and its use in crop improvement; cryo-preservation, Synthetic seed;
11. Introduction to recombinant DNA methods: physical (Gene gun method), chemical (PEG mediated) and Agrobacterium mediated gene transfer methods;
12. Transgenic and its importance in crop improvement;
13. Recombinant DNA technology; Restriction endonucleases: Types and uses; DNA ligases; Vectors: plasmids, cosmids, phagemids, BACs, PACs, YACs, transposon vectors, expression vectors, shuttle vectors, binary plant vectors, co-integrating vectors;
14. Competent cells; Gene isolation and cloning; Genetic transformation of *E. coli*; Gel electrophoresis; Preparation of probes; Southern blotting; Northern blotting; Western blotting;
15. PCR technique and its applications;
16. RFLP, RAPD, SSR;
17. Marker Assisted Breeding in crop improvement;
18. Basics of Genomics, transcriptomics, proteomics and metabolomics;
19. Bioinformatics (Introduction to bioinformatics; Development and scope of bioinformatics; Applications of computers in bioinformatics, Primary and secondary database, Introduction to sequence alignment and its application, Introduction to BLAST and FASTA);
20. Nano-biotechnology (Introduction to nanotechnology; Concepts and Terminology; Nano-Bio Interface; Biological based Nanosystems, molecular motors, biosensors and other devices, DNA nanotechnology);
21. Biotechnology regulations.

Section I

Molecular Biology

Chapter 1

History of Molecular Biology

INTRODUCTION

The study of biology on a molecular level including the structure, function, and organization of biologically important molecules such as DNA, RNA and proteins is referred to as molecular biology. Molecular genetics is the field of molecular biology which deals with the structure and function of genes at the molecular level. Molecular genetics is concerned with the development of biochemical and genetic techniques for handling the complex nucleic acids that constitute genetic material. The major areas of molecular genetics include (i) structure of DNA, (ii) structure of RNA (iii) protein synthesis (iv) gene regulation (v) genetic code, and (vi) use of various molecular techniques. Details of all these techniques are given under respective chapter.

LANDMARK IN THE HISTORY OF MOLECULAR BIOLOGY

Molecular biology was established in the 1930s. However, the term was first coined by Warren Weaver in 1938. The significant milestones in the history of molecular genetics are presented below in the tabular form.

TABLE 1.1: Landmarks in the History of Molecular Genetics

Year	*Scientis(s)*	*Organism*	*Discovery*
1869	Friedrich Miescher	White blood cells	Nucleic Acid
1928	Frederick Griffith	*Diplococcus pneumoniae*	Genetic material
1944	Oswald Theodore Avery, Colin McLeod and Maclyn McCarty	*Diplococcus pneumoniae*	Isolated DNA, called it transforming principle.
1950	Erwin Chargaff	-	In DNA amount of Adenine is equal to amount of Thymine.
1950	Barbara McClintock	Maize	Jumping genes.

Year	*Scientis(s)*	*Organism*	*Discovery*
1952	Hershey and Chase	T2 phage	Proved that DNA is the genetic material.
1953	Watson and Crick	Theoretical	Proposed double helical DNA structure.
1958	Meselson and Stahl	*E. coli*	Proved that DNA replicates in semi-conservative manner.
1961	Jacob and Monod	*E. coli*	Proposed Operon model of gene regulation.
1961-67	Nirenberg, Khorana, Brenner and Crick	-	Crack the genetic code.
1970	Temin and Baltimore	Tumour viruses	Reported Reverse transcription.
1970	Smith et al.	*Haemophilus influenzae*	Discovery of Restriction enzymes.
1972	Walter Fiers and Co-workers	Bacteriophage MS2	First determined sequence of a gene.
1976	Walter Fiers and Co-workers	Bacteriophage MS2	Determined complete nucleotide sequence of bacteriophage.
1977	Sanger, Gilbert and Maxam	Bacteriophage Ö-X174	First sequenced the entire genome of bacteriophage.
1983	Kary Mullis	Basic Research	Discovered Polymerase Chain Reaction [PCR].
1995	Team of Scientists	*Haemophilus influenzae*	The entire genome was sequenced.
2000	Team of Scientists	*Arabidopsis thaliana*	Genome mapping was completed. Genome size is 125 Mb and genes are 25,500.
2002	Team of Scientists	*Rice*	Genome mapping was completed. Genome size is 430 Mb and genes are 56,000.
2005	Syngenta	Rice	A new variety called *Golden Rice 2*.
2005	A Team of Scientists	Rice	Sequenced Rice genome under International Rice. Genome Sequencing Project.
2006	Tuskan *et al.*	Poplar	Sequenced genome of poplar (*Populus tricocarpa*).
2007	Jaillon *et al.*	Grapes	Sequenced genome of Grapes and Papaya.
2008	Ming *et al.*	Apple	Sequenced genome of Apple.
2009	Scnable *et al.*	Maize	Sequenced genome of Maize.
2009	Paterson *et al.*	Sorghum	Sequenced genome of Sorghum.
2010	Schmutz *et al.*	Soybean	Sequenced genome of Soybean.
2010	Chang *et al.*	Castor	Sequenced genome of Castor bean.
2011	Varshney *et al.*	Pigeon pea	Sequenced genome of Pigeon pea.
2011	PGS Consortium	Potato	Sequenced genome of Potato.
2011	Wang *et al.*	Mustard	Sequenced genome of Mustard (*Brassica rapa*).

Year	*Scientis(s)*	*Organism*	*Discovery*
2012	Zheng *et al.*	Fox tail millet	Sequenced genome of Foxtail millet.
2012	Wang *et al.*	Flax	Sequenced genome of Flax.
2012	Wang *et al.*	Cotton	Sequenced genome of Cotton (*Gossypium raimondii*).
2013	Ling *et al.*	Wheat	Sequenced A genome of Wheat (*Triticum urartu*).
2013	Genome Biology.	Tobacco	Sequenced A genome of *Nicotiana sylvestris.*
2014	Kim *et al.*	Pepper	Sequenced Pepper (*Capsicum annuum*) genome.
2015	CSIR	Tulsi	Sequenced Tulsi (*Oscimum sanctum*) genome.
2016	Studer Anthony	Grass	Sequenced genome of grass Diacanthelium.
2016	Bertioli *et al.*	Peanut	Sequenced genome of wild species of peanut [*Arachis ipaensis*]
2017	Jarvis *et al.*	Chenopodium	Sequenced genome of *Chenopodium quinoa.*

INVOLVEMENT OF OTHER DISCIPLINES

Molecular biology is multidisciplinary in nature and involves several disciplines of basic sciences and engineering. It involves various disciplines such as microbiology, biochemistry, genetics, plant physiology, plant pathology, agronomy, physics and electronics and biochemical engineering. The role of each discipline is presented in the Table 1.2.

TABLE 1.2: Involvement of different Disciplines in Plant Biotechnology

Sl.No.	*Name of Discipline*	*Activity in which involved*
1	Microbiology	It is involved in identification of micro-organisms possessing gene of interest such as *Bacillus thuringiensis.*
2	Biochemistry	Biochemistry is involved in the preparation of media for *in-vitro* regeneration of explants.
3	Molecular Genetics	It is involved in isolation, cloning, transfer and integration of gene of interest in desired plants.
4	Genetics	Genetics is involved for *in-vitro* characterization of plantlets for various genetic traits.
5	Plant Physiology	Plant physiology is involved in embryo culture, identification of plantlets with resistance to drought, salinity, metal toxicity, heat, frost, *etc.*
6	Plant Pathology	It is involved in sterilization of explants and keeping the tissue culture laboratory pathogen free.
7	Agronomy	It is involved in hardening and acclimatization of *in-vitro* derived plants under field conditions.

Sl.No.	Name of Discipline	Activity in which involved
8	Physics and Electronics	It is involved in designing various instruments and equipment required for use in plant biotechnology.
9	Biochemical Engineering	Biochemical engineering is involved in mass production of various bio-chemicals such as alcohol, pharmaceuticals, *etc.*

APPLICATIONS

There are several applications of Molecular Genetics in the field of crop improvement, biotechnology, medical science and evolution. These are briefly discussed below:

(i) **Crop Improvement:** Molecular genetics plays an important role in the genetic improvement of crop plants. This can be achieved in two ways, *viz.* (i) by developing transgenic plants, and (ii) by developing mutant varieties of crop plants. These techniques can be used for developing genotypes superior in resistance to insects, diseases, quality, productivity or any other characteristics.

(ii) **Biotechnology**: Similarly, the application of molecular genetics has completely transformed the field of biotechnology, with new possibilities ranging from the treatment of human diseases to the development of new forms of crops. Molecular genetics is not only the most rapidly developing biological science of this decade, but is also the most promising and exciting science of the next few decades.

(iii) **Medical Science**: Medical research is one field in which Molecular Genetics plays a vital role. The application of molecular genetics has solved problems in many areas of biological, biochemical and medical research. It is used to study human gene therapy, and investigate such things as the molecular basis of cancer, cell growth and development, and diseases like AIDS. In future, it is expected to result in exciting new discoveries in medical science. Molecular genetics has helped in genome sequencing in human and several other organisms. Now genomic sequences can be analyzed, dissected, recombined and reproduced in ways that formerly appeared impossible.

Researchers have already identified single genes associated with a number of diseases, such as cystic fibrosis. As research progresses, investigators will also uncover the mechanisms for diseases caused by several genes or by single genes interacting with environmental factors. Genetic susceptibilities have been implicated in many major disabling and fatal diseases including heart disease, stroke, diabetes, and several kinds of cancer. The identification of these genes and their proteins will pave the way to more effective therapies and preventive measures.

(iv) **Evolution:** Molecular genetics has useful applications in the study of the process of organic evolution. It is used for the study of the molecular bases of evolution of both plants and animals. Investigators determining the underlying biology of genome organization and gene regulation will also begin to understand how humans develop, why this process sometimes goes wrong, and what changes take place as people age.

SUMMARY

The study of biology on a molecular level including the structure, function, and organization of biologically important molecules such as DNA, RNA and proteins is referred to as molecular biology. Molecular genetics is the field of molecular biology which deals with the structure and function of genes at the molecular level. Molecular genetics is concerned with the development of biochemical and genetic techniques for handling the complex nucleic acids that constitute genetic material. The major areas of molecular genetics include (i) structure of DNA, (ii) structure of RNA, (iii) protein synthesis, (iv) gene regulation, (v) genetic code, and (vi) use of various molecular techniques.

Molecular biology is multidisciplinary in nature and involves several disciplines of basic sciences and engineering. It involves various disciplines such as microbiology, biochemistry, genetics, plant physiology, plant pathology, agronomy, physics and electronics and biochemical engineering. The role of each discipline has been explained. Important events in the history of molecular biology are presented in tabular form.

There are several applications of Molecular Genetics in the field of crop improvement, biotechnology, medical science and evolution. In medical science, for example, researchers have already identified single genes associated with a number of diseases, such as cystic fibrosis. In biotechnology it has wide applications ranging from treatment of human diseases to the development of new forms of crops. It is used for the study of the molecular bases of evolution of both plants and animals.

GLOSSARY

DNA fingerprinting: A technique that is used to distinguish between individuals of the same species using only samples of their DNA; also called genetic fingerprinting, DNA testing, DNA typing, or DNA profiling.

DNA Micro Arrays: A collection of spots attached to a solid support such as a microscope slide where each spot contains one or more single-stranded DNA.

DNA Sequencing: The process of determining the order of the nucleotide bases along a DNA strand

Expression Cloning: One of the most basic techniques of molecular biology to study protein function.

Gel electrophoresis: One of the principal tools of molecular biology. The basic principle is that DNA, RNA, and proteins can all be separated by means of an electric field.

Gene Cloning: The isolation of a specific fragment of target (*e.g.* human) DNA by propagating that fragment in a microorganism (usually a bacterium but sometimes a yeast).

Molecular Biology: The study of biology on a molecular level including the structure, function, and organization of biologically important molecules such as DNA, RNA and proteins.

Molecular Genetics: It is the study of structure and function of genes at the molecular level.

Northern Blot Technique: A method of detecting specific type of RNA molecule among of a set of different samples of RNA.

Polymerase Chain Reaction [PCR]: An extremely versatile technique for copying DNA.

Southern Blot Technique: A technique of detecting a small number of DNA fragments which may be present in a complex mixture.

Western Blot Technique: A technique of detecting specific protein in a sample.

QUESTIONS

1. **Describe the role of various disciplines in the study of molecular biology.**
2. **What is molecular genetics? Describe briefly it applications.**
3. **Give in brief the contribution of the following Scientists:**
   (i) Watson and Crick
   (ii) Kary Mullis
   (iii) Temin and Baltimore
   (iv) Jacob and Monod
   (v) Edwin Southern
   (vi) Patricia Thomas
4. **Explain briefly ten significant milestones in the history of molecular genetics.**
5. **Describe in brief the role of molecular genetics in crop improvement.**
6. **Describe in brief the role of molecular genetics in Medical Science.**
7. **Explain role of molecular genetics in evolution.**

*****

Chapter 2

DNA Structure and Function

INTRODUCTION

DNA refers to the molecules inside cells that carry genetic information and pass it from one generation to the next. In other words, the material in a cell that contains unit of heredity is referred to as genetic material. Genetic material stores the genetic information of an individual. The scientific name for DNA is deoxyribonucleic acid. In most of the living organisms, DNA (deoxyribonucleic acid) is the genetic material. However, in some viruses, RNA (ribonucleic acid) is the genetic material.

MAIN FEATURES

Mostly the DNA structure is double stranded in both eukaryotes and prokaryotes. However, in some viruses DNA is single stranded. DNA is a double-stranded molecule held together by weak bonds between base pairs of nucleotides. The four nucleotides in DNA contain the bases: adenine (A), guanine (G), cytosine (C), and thymine (T). In eukaryotes, the DNA is of linear shape.

In prokaryotes and mitochondria, the DNA is circular. In eukaryotes, DNA is found both in nucleus and cytoplasm. In the nucleus it is a major component of chromosome, whereas in cytoplasm it is found in mitochondria and chloroplasts. In prokaryotes, it is found in the cytoplasm.

The DNA is capable of self-replication. This is the only chemical which has self-replicating capacity. The DNA replicates in semi-conservative manner. In eukaryotes, DNA replicates during the S-Phase of the cell cycle.

There are different forms of DNA such as A, B, C, D and Z DNA. The DNA may be linear or circular. It may be double stranded or single stranded. It may be right handed or left handed. The DNA may be repetitive or unique. It may be nuclear or cytoplasmic.DNA plays important role in various ways. It is used in transcription, *i.e.*, synthesis of mRNA which in turn is used in protein synthesis.It carries genetic information and passes it from one generation to the next generation. DNA stores and transmits the genetic information in cells. It forms the basis for genetic code. The genes are made of DNA and are responsible for passing on traits from generation to generation. DNA contains the genetic instructions for the development and functioning of living organisms. Thus it is the substance of heredity.

DNA AS GENETIC MATERIAL

There are some experimental evidences which prove that in most of the organisms DNA is the genetic material and only in some viruses RNA is the genetic material. These evidences include, (1) transformation, (2) bacteriophage infection, (3) transduction, and (4) biochemical analysis. All these are briefly discussed as follows:

1. Transformation

Transformation refers to the genetic recombination in which naked DNA from one cell can enter and integrate in another cell. Experiments on genetic transformation were conducted by Griffith (1920) and Avery, Macleod and McCarty (1944) which support that DNA is the genetic material.

Griffith (1920) working with pneumonia causing bacteria *Diplococcus pneumoniae* suggested that the chemical from the heat killed bacteria transformed the avirulent bacteria into virulent which caused death of the mice. However, Griffith could not conclude that the transforming chemical was DNA.

Avery, Macleod and McCarty [1944] working with pneumonia causing bacteria *Diplococcus pneumoniae* concluded that DNA was the genetic material and not the proteins. Because the transformation occurred when DNA was present in the extract and there was no transformation when DNA was digested with DNAse enzyme.

2. Bacteriophage Infection

Bacteriophage infection provides evidence in support of DNA as the genetic material. Bacteriophages or phages are viruses which attack bacteria and kill them. The phage has two parts, *viz.* tail and head. The tail is composed of protein only. The head has outer coat of protein and inner core as DNA. The phage infects bacteria by attaching its tail to the body of the bacteria. The DNA enters the bacterial cell and multiplies inside and protein portion remains outside the bacterial cell. After replication, phages come out of the bacterial cell and make the outer coating their own.

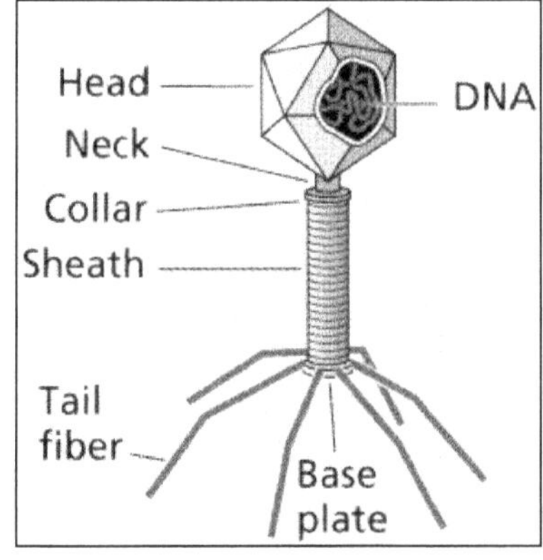

FIGURE 2.1: Structure of a Bacteriophage Virus.

Hershey and Chase [1951] working with T2 phage of *E. coli* observed that only the phages which were labelled with radio-active phosphorus [P32] transmitted radio activity to their daughter phages. This provided a strong evidence that DNA is the genetic material and not the protein. They concluded that the genetic material must be DNA, carrying the radioactivity inside the bacteria and to offspring phages.

3. Transduction

The genetic recombination in bacteria in which DNA is transferred from one bacterial cell to another via the bacteriophage is called transduction. Bacteriophages attack the bacteria, lyse them and multiply inside the bacteria. In this process, sometimes a small DNA segment of bacteria is contained by the bacteriophage. When the bacteriophage attacks another bacterium, the DNA of previous bacteria from bacteriophage is integrated with the DNA of new bacteria and changes some of the features of new bacteria. Transduction is of two types, *viz.* general and specialized.

- **(i) General Transduction:** When any part of donor DNA is transferred to the recipient bacteria, it is known as general or unrestricted transduction. When a segment of donor DNA replaces a corresponding segment of host DNA, it is called complete transduction. Sometimes donor DNA fails to recombine with host DNA, it is known as abortive transduction.
- **(ii) Specialized Transduction:** When bacteriophage is able to transduct only a restricted portion of host DNA, it is referred to as restricted or specialized transduction. Transduction has been reported in several bacteria such as *E. coli, Pseudomonas, Salmonella, etc.*

4. Biochemical Evidences

There are several biochemical evidences which support that DNA is the genetic material. Biochemical estimation of DNA in various tissues and ploidy levels provides additional support for DNA as the genetic material. Some such evidences are briefly presented below:

- (i) The DNA is predominantly present in the chromosomes. This also suggests that DNA is the genetic material.
- (ii) The germ cells of higher plants and animals are haploid with only one set of chromosomes. Such cells contain half of the amount of DNA present in the somatic cells of the same species.
- (iii) The amount of DNA in the somatic tissues differs from species to species, but is constant for a species and cannot be altered by change either in the external or internal environment of the organism. The amount of DNA is also constant per diploid cell of an organism [Alfred Mirsky].
- (iv) The amounts of each of the four nucleotide bases vary considerably from species to species, but the DNA of any given species contains equal amounts of adenine and thymine and equal amounts of cytosine and guanine [Chargaff].
- (v) The amount of DNA in a species varies as per the ploidy level. The tetraploid will contain double the amount of DNA than diploid species.
- (vi) Some amount of DNA is also found in the cytoplasm especially in the chloroplasts and mitochondria. Such DNA controls the inheritance of cytoplasmic genes, because some characters exhibit cytoplasmic inheritance.

MOLECULAR STRUCTURE OF DNA

The double helical structure of DNA was proposed by Watson and Crick in 1953. This work resulted in significant breakthrough in understanding the gene function. This structure has been verified in many different ways and is universally accepted. James Watson and Francis Crick were awarded Nobel Prize in 1958 for this significant contribution in the field of molecular biology. The important features of their model of DNA are as follows:

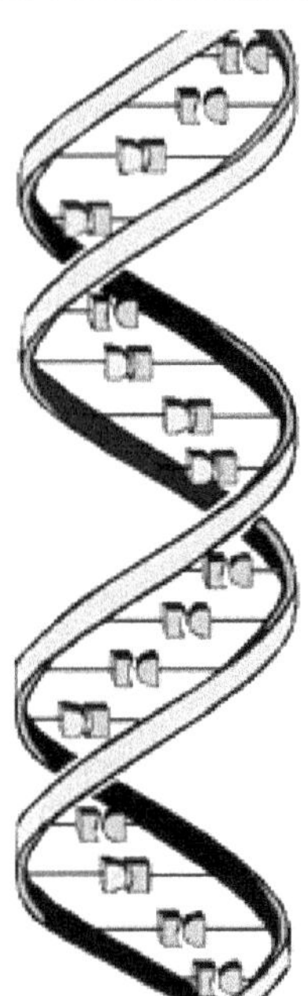

FIGURE 2.2: Molecular Structure of DNA.

(i) Two helical polynucleotide chains are coiled around a common axis, the chains run in opposite directions.

(ii) The two chains held together by hydrogen bonds formed between pairs of bases. Pairing is highly specific. Adenine pairs with thymine, guanine always pairs with cytosine. A = T, G = C.

(iii) The purine and pyrimidine bases are on the inside of the helix whereas the phosphate and deoxyribose units are situated on the outside. Also the planes of the base residues are perpendicular to the helix axis. While the planes of the sugar residues are almost at right angles to those of the bases.

(iv) The diameter of the helix is 2 nm. Adjacent bases are separated by 0.34 nm along the helix axis. Hence the helix repeats itself every 10 residues on each chain at intervals of 3.4 nm.

(v) The sequence of bases along the polynucleotide chain is not restricted. The precise sequence of bases carries the genetic information.

(vi) The sugar-phosphate backbones of the two DNA strands wind around the helix axis like the railing of a spiral staircase.

(vii) The bases of the individual nucleotides are on the inside of the helix, stacked on top of each other like the steps of a spiral staircase.

COMPONENTS OF DNA

DNA molecule is a polymer which is composed of several thousand pairs of nucleotide monomers. Union of several nucleotides together leads to the formation of polynucleotide chain. The monomer units of DNA are nucleotides, and the polymer is known as a "polynucleotide." Each nucleotide consists of a 5-carbon sugar (deoxyribose), a nitrogen containing base attached to the sugar, and a phosphate group. Thus there are three components of DNA, *viz.* (1) nitrogenous bases, (2) deoxyribose sugar, and (3) phosphate group. These are briefly discussed below:

1. Nitrogenous Bases

Nucleotides are also known as nitrogenous bases or DNA bases. Nitrogenous base are of two types, *viz.* pyrimidines and purines. Pyrimidines are single ring structures. These are of two types namely cytosine and thymine.They occupy less space in DNA structure. Pyrimidine is linked with deoxyribose sugar at position 3.

Purines are double ring compounds. They are of two types, *viz.* adenine and guanine. They occupy more space in DNA structure. Deoxyribose sugar is linked at position 9 of purine.

Thus, in DNA there are four different types of nitrogenous bases, *viz.* adenine (A), guanine (G), cytosine (C) and thymine (T). In RNA, the pyrimidine base thymine is replaced by uracil.

Base Pairing

The purine and pyrimidine bases always pair in a definite fashion. Adenine will always pair with thymine and guanine with cytosine. Adenine and thymine are joined by double hydrogen bonds while guanine and cytosine are joined by triple hydrogen bonds. However, these bonds are weak which help in separation of DNA strands during replication.

2. Deoxyribose Sugar

This is a pentose sugar having five carbon atoms. The four carbon atoms are inside the ring and the fifth one is with CH_2 group. This has three OH groups on 1,3 and 5 carbon positions. Hydrogen atoms are attached to carbon atoms one to four. In RNA, the sugar is ribose is similar to deoxyribose except that it has OH group on carbon atom 2 instead of H group.

3. Phosphate

The phosphate molecule is arranged in an alternate manner to deoxyribose molecule. Thus there is deoxyribose on both sides of phosphate. The phosphate is joined with carbon atom 3 of deoxyribose at one side and with carbon atom 5 of deoxyribose on the other side.

Nucleosides and Nucleotides

A combination of deoxyribose sugar and nitrogenous base is known as nucleoside and a combination of nucleoside and phosphate is called nucleotide.

Nucleoside = Deoxyribose Sugar + Nitrogenous base

Nucleotide = Deoxyribose + Nitrogenous base + Phosphate

Thus, a nucleotide is a nucleoside with one or more phosphate groups covalently attached to it. Nucleosides differ from nucleotides in that they lack phosphate groups. The four different nucleosides of DNA are deoxyadenosine (dA), deoxyguanosine (dG), deoxycytosine (dC), and (deoxy)thymidine (dT, or T) DNA Backbone

The DNA backbone is a polymer with an alternating sugar-phosphate sequence. The deoxyribose sugars are joined at both the 3′-hydroxyl and 5′-hydroxyl groups to phosphate groups in ester links, also known as "phosphodiester" bonds.

FORMS OF DNA

Depending upon the nucleotide base per turn of the helix, pitch of the helix, tilt of the base pair and humidity of the sample, the DNA can be observed in four different forms namely, A, B, C and D. A comparison of A, B and Z forms of DNA is presented in Table 2.1.

B-Form

This is the same form of DNA proposed by Watson and Crick. This is the most common form of DNA. It is observed when humidity is 92 per cent and salt concentration is high. The coiling is in the right direction and the number of base is 10 per turn of helix. The pitch is 3.4 nm and the sugar phosphate linkage is normal. The helix is narrower and more elongated than A form. The major groove is wide which is easily accessible to proteins. The minor groove is narrow. The conformation is favored at high water concentrations. The base pairing is nearly perpendicular to helix axis and the sugar puckering is $C_{2'}$-endo.

A-Form

This form is observed when the humidity of the sample is 75 per cent. The coiling is in the right direction and the number of bases is 10.7 per turn of helix. The pitch is 2.8 nm and the sugar phosphate linkage is normal. The major groove is deep and narrow which is not easily accessible to proteins. The minor groove is wide and shallow which is accessible to proteins, but information content is lower than major groove. The helix is shorter and wider than B form. The conformation is favored at low water concentrations. The base pairs are tilted to helix axis and the sugar puckering is $C_{3'}$-endo.

Z-Form

The helix has left-handed coiling pattern and the number of bases is 12 per turn of helix. The pitch is 4.5 nm and the sugar phosphate linkage is zig-zag. The major "groove" is not really a groove and the minor groove is narrow. The helix is narrower and more elongated than A or B form. The base pairing is nearly perpendicular to helix axis. The conformation is favored by high salt concentrations and the sugar puckering is C : $C_{2'}$-end, G : $C_{2'}$-endo.

Denaturation and Renaturation

The process of breaking of hydrogen bonds between the strands and separation of two strands on heating of DNA molecule is called denaturation. The DNA in disrupted state is known as denatured. Denaturation occurs at 100 degree Celsius. The process by which denatured DNA is changed in to original DNA is called renaturation and reformed DNA is known as renatured DNA. The cooling of denatured DNA gradually leads to renaturation. The optimum temperature for renaturation is 20-25 degree Celsius. The denaturation and renaturation ability of DNA is useful for developing heteroduplexes.

CLASSIFICATION OF DNA

The DNA can be classified into various types depending upon number of strand, position, shape, coiling pattern, copies per genome, nucleotide per turn of base as presented in the Table 2.1.

TABLE 2.1: Classification of DNA

Sl. No.	*Basis of Classification*	*Types of DNA*	*Explanation/Brief Description*	*Example/ Remarks*
1	Number of strand	Double stranded	Have two strands spirally arranged.	Eukaryotes
		Single stranded	Have single strand	Prokaryotes
2	Position	Nuclear DNA	Found in nucleus in chromosomes.	Eukaryotes
		Cytoplasmic DNA	Found in protoplast and Mitochondria in cytoplasm.	Both prokaryotes and eukaryotes
3	Shape	Linear DNA	Have straight or linear shape	Eukaryotes
		Circular DNA	Have circular shape	Prokaryotes and cytoplasm
4	Copy per genome	Unique DNA	Have single copy per genome.	Eukaryotes
		Repetitive DNA	Have small sequences of bases repeated several hundred times.	Eukaryotes except Yeast.
5	Coiling pattern	Right handed DNA	Have coiling in right direction.	Eukaryotes
		Left handed DNA	Have coiling in left direction.	Prokaryotes
6	Nucleotides per turn of base	A-Form DNA	Have 10.7 nucleotides per turn of base.	Right handed coiling
		B-Form DNA	Have 10 nucleotides per turn of base.	Right handed coiling
		Z-Form DNA	Have 12 nucleotides per turn of base.	Left handed coiling

FUNCTIONS OF DNA

There are five important functions of DNA, *viz.* RNA synthesis, protein synthesis, inheritance, control of metabolic pathways and evolution. These are explained as follows:

1. The synthesis of RNA takes place using DNA as template which is then used in protein synthesis.
2. DNA control protein synthesis by synthesizing three types of RNAs, *viz.* m-RNA, t-RNA and r-RNA.
3. Characters are inherited from parents to their offspring through genes which are composed of DNA.
4. DNA controls the enzymes which in turn control the metabolic pathway of an individual.
5. DNA plays important role in the organic evolution through gene mutations. Mutations give rise to new phenotypes of an individual.

SUMMARY

DNA refers to the molecules inside cells that carry genetic information and pass it from one generation to the next. A nucleic acid that carries the genetic information in the cell and is capable of self-replication and RNA synthesis is referred to as DNA. The scientific name for DNA is deoxyribonucleic acid.

The molecular double helical structure of DNA was proposed by Watson and Crick in 1953, which is universally accepted. James Watson and Francis Crick were awarded Nobel Prize in 1958 for this significant contribution in the field of molecular biology.

DNA is a polymer. The monomer units of DNA are nucleotides, and the polymer is known as a "polynucleotide." There are three components of DNA, *viz.* (1) nitrogenous bases, (2) deoxy-ribose sugar, and (3) phosphate group. These have been discussed. Depending upon the nucleotide base per turn of the helix, pitch of the helix, tilt of the base pair and humidity of the sample, the DNA can be observed in four different forms namely, A, B, C and D. The B form is the most common form and is the same as proposed by Watson and Crick. Other forms of DNA are rare forms. A comparison of A, B and Z forms of DNA is presented in Tabular form. The DNA can be classified into various types depending upon number of strand, position, shape, coiling pattern, copies per genome, nucleotide per turn of base. There are five important functions of DNA, *viz.* RNA synthesis, protein synthesis, inheritance, control of metabolic pathways and evolution.

GLOSSARY

Base-pairing: The pairing of purine and pyrimidine bases in DNA molecule.

Circular DNA: DNA that has a ring or circular shape. Such DNA is usually found in prokaryotes, chloroplasts and mitochondria.

Cytoplasmic DNA: DNA which is found in the cytoplasm in chloroplasts and mitochondria.

DNA: A nucleic acid that carries the genetic information in the cell and is capable of self-replication and RNA synthesis.

Deoxyribose: A pentose sugar having five carbon atoms.

Double Stranded DNA: DNA having spirally arranged two strands.

Left handed DNA: DNA with coiling of the helix in the left direction such as Z-DNA.

Linear DNA: DNA that has thread-like structure with both ends free. Such DNA is found in eukaryotes.

Nuclear DNA: DNA which is found in the nucleus in the chromosomes.

Nucleoside: A combination of deoxyribose sugar and nitrogen base.

Nucleotide: A combination of deoxyribose sugar, nitrogen base and phosphate linkages.

Promiscuous DNA: The DNA segments which move from one cell organelle to another, *i.e.*, from chloroplasts to mitochondria and nucleus.

Pyrimidines: Nitrogenous base with single ring structure.

Purines: Nitrogenous base with double ring structure.

Repetitive DNA: DNA in which small sequences of bases are repeated several hundred times. Such DNA is found in all eukaryotes except yeast.

Right handed DNA: DNA with coiling of the helix in the right direction, a usual form of DNA.

Single Stranded DNA: DNA having one helix only. Found in some bacteriophage.

Strand: Each of the two helix of DNA.

Transformation: The genetic recombination in which naked DNA from one cell can enter and integrate in another cell.

Transduction: The genetic recombination in bacteria in which DNA is transferred from one bacterial cell to another via the bacteriophage.

Unique DNA: The DNA segment having single copy per genome.

Watson and Crick Base Pairing: Normal base pairing, *viz.* A with T and G with C in DNA and A with U and G with C in RNA.

Z-DNA: The DNA in which sugar and phosphate linkages follow a zig-zag pattern. Such DNA has left handed double helical structure.

QUESTIONS

1. **How would you prove that DNA is the genetic material?**
2. **Define DNA and explain important features of B form of DNA.**
3. **What is Z-DNA? Describe its important features.**
4. **Describe important features of A form of DNA molecule.**
5. **Give brief comparison of A, B and Z forms of DNA. Which is the most common form of DNA?**
6. **Give a comparison of eukaryotic and prokaryotic DNA.**
7. **What are different components of DNA structure? Describe them briefly.**
8. **Describe briefly Watson and Crick model of DNA structure.**
9. **Differentiate between the following:**

   (i) Nucleoside and nucleotide (ii) Purines and Pyrimidines

   (iii) A-DNA and Z-DNA (iv) Linear DNA and circular DNA
10. **Write short notes on the following:**

    (i) Repetitive DNA (ii) Unique DNA

    (iii) Promiscuous DNA (iv) Cytoplasmic DNA

11. Define nucleotide and describe pairing pattern of nucleotides in DNA.

12. Write short notes on the following:

(i) Transformation (ii) Transduction

(iii) Denaturation (iv) Renaturation

*****

Chapter 3

DNA Replication

INTRODUCTION

The process by which a DNA molecule makes its identical copies is referred to as DNA replication. In other words, it is the process of duplicating the DNA to make two identical copies. This process is important in all known forms of life and the general mechanisms of DNA replication are the same in prokaryotic and eukaryotic organisms. The main points related to DNA replication are briefly presented below.

1. The DNA replication takes place during S substage of interphase.
2. In eukaryotes, replication occurs in the cell nucleus, whereas in prokaryotes it occurs in the cytoplasm.
3. The existing DNA is used as a template for the synthesis of new DNA strands. It is possible that during replication on strand of DNA can replicate continuously or the other discontinuously or in piece. The continuously replicating strand is known as leading strand and the discontinuously replicating strand is known as lagging strand.
4. The process of DNA replication takes place under the control of DNA polymerase.
5. The synthesis of one new strand takes place in 5-3 and that of other in opposite (3-5) direction.
6. Based on the direction, the replication may be unidirectional or bidirectional. On the basis of continuity, the replication may be continuous or discontinuous.
7. The point of initiation of DNA replication is known as origin. The progress of replication process is measured from the point of origin.

MODELS OF DNA REPLICATION

There are three models which explain the accurate replication of DNA. These are: (i) dispersive replication, (ii) conservative replication, and (iii) semi- conservative replication (Figure 3.1). These are explained as follows:

1. **Conservative Replication:** According to this model of DNA replication two DNA molecules are formed from parental DNA. One copy has both parental strands and the other contains both newly synthesized strands (Figure 3.1a). This method is also not accepted as there is no experimental proof in support of this model.
2. **Semi-conservative Replication:** This model of DNA replication was proposed by Watson and Crick. According to this model of DNA replication, both strands of parental DNA separate from each other. Each old strand synthesizes a new strand. Thus each of the two resulting DNA molecules has one parental and one new strand. This model of DNA replication is universally accepted because there are several evidences in support of this model.
3. **Dispersive Replication:** According to this model of replication the two strands of parental DNA break at several points resulting in several pieces of DNA. Each piece replicates and pieces are reunited randomly resulting in formation of two copies of DNA from single copy. The new

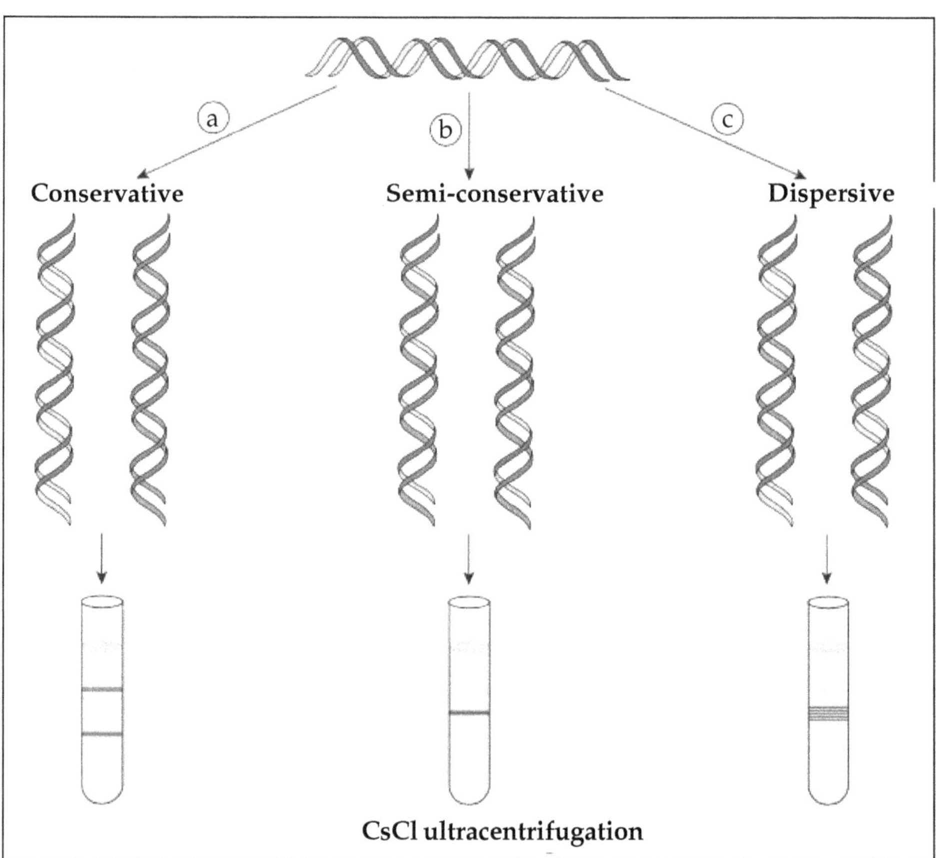

FIGURE 3.1a, b & c: Models of DNA Replication.

DNA molecules are hybrids which have new and DNA in patches (Figure 3.1 c). This method of DNA replication is not accepted as it could not be proved experimentally.

STEPS IN DNA REPLICATION [EUKARYOTES]

The semi-conservative model of DNA replication consists of six important steps, *viz.* (1) unwinding, (2) binding of RNA primase, (3) elongation, (4) removal of primers, (5) termination, and (6) DNA repair. These are briefly discussed as follows:

1. **Unwinding**: The first major step in the process of DNA Replication is the breaking of hydrogen bonds between bases of the two antiparallel strands. The unwinding of the two strands is the starting point. The splitting happens in places of the chains which are rich in A-T. That is because there are only two bonds between Adenine and Thymine, whereas there are three hydrogen bonds between Cytosine and Guanine. The **Helicase** enzyme splits the two strands. The initiation point where the splitting starts is called origin of replication. The structure that is created is known as Replication Fork.

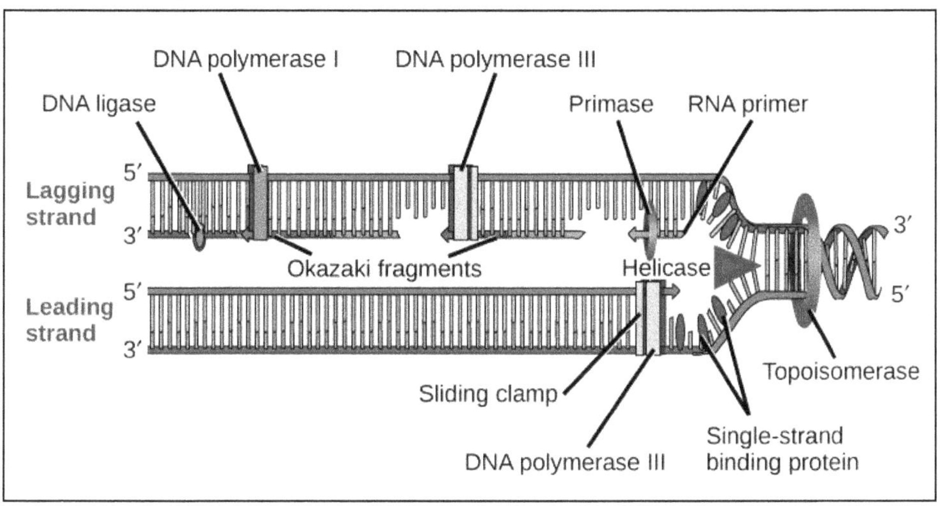

FIGURE 3.2: Mechanism of DNA Replication.

2. **Binding of RNA Primase**: Synthesis of RNA primer is essential for initiation of DNA replication. RNA primer is synthesized by DNA template near the origin with the help of RNA Primase. RNA Primase can attract RNA nucleotides which bind to the DNA nucleotides of the 3′-5′ strand due to the hydrogen bonds between the bases. RNA nucleotides are the primers (starters) for the binding of DNA nucleotides.
3. **Elongation**: The elongation proceeds in both directions, *viz.* 5′-3′ and 3′-5′ template. The 3′-5′ proceeding daughter strand - that uses a 5′-3′ template - is called leading strand because DNA Polymerase a can "read" the template and continuously add nucleotides. The 3′-5′ template cannot

be "read" by DNA Polymerase a. The replication of this template is complicated and the new strand is called lagging strand. In the lagging strand the RNA Primase adds more RNA Primers. DNA polymerase a reads the template and lengthens the bubbles. The gap between two RNA primers is called "Okazaki Fragments. The RNA Primers are necessary for DNA Polymerase a to bind Nucleotides to the 3′ end of them. The daughter strand is elongated with the binding of more DNA nucleotides.

4. **Removal of Primers:** The RNA Primers are removed or degraded by DNA polymerase I. This enzyme also catalyzes the synthesis of short DNA segments to replace the primers. The gaps are filled with the action of DNA Polymerase which adds complementary nucleotides to the gaps. The DNA Ligase enzyme adds phosphate in the remaining gaps of the phosphate-sugar backbone. Each new double helix is consisted of one old and one new chain. This is called semi-conservative replication.
5. **Termination**: The termination takes place when the DNA Polymerase reaches to an end of the strands. In other words, it is the separation of replicated linear DNA. After removal of the RNA primer, it is not possible for the DNA Polymerase to seal the gap because there is no primer. Hence, the end of the parental strand where the last primer binds is not replicated. These ends of linear (chromosomal) DNA consist of noncoding DNA that contains repeat sequences and are called telomeres. A part of the telomere is removed in every cycle of DNA Replication.
6. **DNA Repair:** The DNA Replication is not completed without DNA repair. The possible errors caused during the DNA replication are repaired by DNA repair mechanism. Enzymes like nucleases remove the wrong nucleotides and the DNA polymerase fills the gaps. Similar processes also happen during the steps of DNA replication of prokaryotes though there are some differences.

REPLICATION IN PROKARYOTES

In prokaryotes [bacteria and viruses], the DNA is circular. Hence, replication of DNA in these organisms differs from that of eukaryotes, where DNA is linear. Several model of replication of circular DNA have been suggested. There are two most commonly known models of circular DNA replication, *viz.* (1) Cairns Model, and (2) rolling circle model. There are briefly discussed as follows:

1. Cairns Model of DNA Replication

This model of DNA replication in prokaryotes was proposed by Cairns in 1963. It explains thc mechanism of DNA replication in double stranded circular DNA of bacteria. According to this model the DNA replication consists of following important steps.

(i) **Unwinding of DNA**: The double stranded circular DNA starts unwinding or separation at a specific point called origin. Two growing points are established. As the growing points move apart, unwinding of the DNA double strand takes place. This unwinding creates torque since the parental

DNA strands cannot unwind freely.

(ii) **Initiation of Replication**: At the point of origin, bidirectional replication is initiated. Both strands of DNA are replicated. This leads to further unwinding of DNA double strand, resulting in formation of torque.

(iii) **Super Twisting:** The torque leads to super twisting of DNA strand. As a result of super twisting, one of the strands is cut (nicked) which makes the parental strand to rotate freely. The cut is made by a swiveling protein, which relieves the strain.

(iv) **Sealing of Broken Points:** The breaks are sealed by swiveling protein and thus the replication is over. The replication process continues in this way.

Cairns type replication has been demonstrated in the bacteria *E. coli* and *Bacillus subtilis*, in several viral and plasmid chromosomes and in DNA synthesis of mitochondria and chloroplasts.

2. Rolling Circle Model of DNA Replication

This model of circular DNA replication was proposed in 1968 [Gilbert and Dressler,1968; Eisen, Pareira da Silva and Jacob, 1968]. This model explains mechanism of DNA replication in single stranded circular DNA of viruses, *e.g.* ϕX174, and the transfer of *E. coli* sex factor (plasmid). The ϕX174 chromosome consists of a single stranded DNA ring (Positive Strand). This model is most widely accepted. The mechanism of replication consists of following important steps:

(i) **Synthesis of New Strand**: First the chromosome becomes double stranded by synthesis of a negative strand. The original strand is positive. The negative strand is synthesized in side of parental positive strand.

(ii) **Cut in Outer Strand**: The negative or inner strand remains a close circle and the positive strand is nicked at a specific site by endonuclease enzyme. This enzyme recognizes a particular sequence at this point. Thus a linear strand with 3 and 5 ends is created.

(iii) **Formation of Tail:** The original positive strand comes out in the form of a tail of a single linear strand as a consequence of rolling circle. The 5′ end of the broken strand becomes attached to the plasma membrane of the host bacterium. Such replicating phage DNA is commonly found associated with bacterial membranes. The unbroken parental strand rolls and unwinds as synthesis proceeds, leaving a 'tail' which is attached to the membrane.

(iv) **Synthesis of New Strand**: The synthesis of new strand takes place along the parental strand at the tail end in a 3-5 direction. The 3′ end serves as a primer for the synthesis of a new DNA strand under the catalytic action of DNA polymerase. The unbroken strand is used as the template for this purpose, and a complementary strand is synthesized. Thus the parental molecule itself is used as a primer for initiating replication.

New DNA is also synthesized in the tail region in discontinuous segments in the 5-3 direction. This synthesis is presumably preceded by the synthesis of an RNA primer under the catalytic action of RNA polymerase. The tail is cut off by a specific endonuclease into a unit length progeny rod.

(v) **Cutting of Tail:** Now the tail is cut off into a linear segment by endonuclease. The linear segment becomes circular by joining two ends with the help of ligase enzyme. Thus a new circular molecule is formed which can become new rolling circle and replicate further.

Genetic information is preserved in the single stranded template ring which remains circular and serves as an endless template. There is no swiverling problem or creation of torque in the rolling circle model. As the strands unwind the 3′ end is free to rotate on the unbroken strand. The growing point itself thus serves as a swivel.

Evidence for the rolling circle model has been obtained from the replication of several viruses (M13, P2, T4, dikudd), replication resulting in transfer of genetic material during mating of bacteria, and the special DNA synthesis during oogenesis in Xenopus.

TABLE 3.1: Comparison of DNA Replication in Eukaryotes and Prokaryotes

Sl.No.	*Particulars*	*Eukaryotes*	*Prokaryotes*
1	Time of replication	S phase	End of cell cycle
2	Site of replication	Nucleus	Cytoplasm
3	Template used	Pre existing DNA	Pre existing DNA
4	Enzymes involved	DNA polymerase alpha, delta, gamma and epsilon	DNA polymerase I, II and III
5	Direction of replication	Both 3-5 and 5-3	Both 3-5 and 5-3
6	Replication model	Semi-conservative	Semi-conservative
7	Mechanism	More complex	Simple
8	Active replication sites	Many	One
9	Replication rate per second	50 bases	500 bases
10	Length of Okazaki fragments	1500 bases	150 bases

SUMMARY

The process by which a DNA molecule makes its identical copies is referred to as DNA replication. In other words, it is the process of duplicating the DNA to make two identical copies. The DNA replication has been discussed in relation to time of replication, site of replication, template used, enzymes involved, direction of replication, origin of replication, rate of replication, type of replication and model of replication.

There are three models which explain the accurate replication of DNA. These are: (i) conservative replication, (ii) semi- conservative replication, and (iii) dispersive replication.

The semi-conservative model [mechanism] of DNA replication consists of six important steps, *viz.* (1) unwinding, (2) binding of RNA primase, (3) elongation, (4) removal of primers, (5) termination, and (6) DNA repair.

In prokaryotes [bacteria and viruses], the DNA is circular. Hence, replication of DNA in these organisms differs from that of eukaryotes, where DNA is linear. Several model of replication of circular DNA have been suggested. There are two most commonly known models of circular DNA replication, *viz.* (1) Cairns Model, and (2) rolling circle model. These have been discussed.

GLOSSARY

Bidirectional Replication: Replication of DNA in both the directions from the point of origin.

Conservative Replication: A model of DNA replication in which one new DNA has parental strands and other contains both newly synthesized strands.

Continuous Replication: Uninterrupted replication of DNA from one end to other end.

Discontinuous Replication: DNA replication in pieces from one end to other end.

Dispersive Replication: A model of DNA replication in which the new DNA molecules have old and new DNA in patches.

DNA Replication: The process by which a DNA molecule makes its identical copies.

Lagging Strand: The discontinuously replicating strand of parental DNA molecule.

Leading Strand: The continuously replicating strand of parental DNA molecule.

Non-sense Strand: The DNA strand which is not used for synthesis of RNA.

Okazaki Fragments: Short segments of nucleotides synthesized in in lagging strand of DNA as a result of discontinuous replication.

Origin: The point of initiation of DNA replication.

Semi-conservative Replication: A model of DNA replication in which each of two new DNA molecule contains one old and one new strand.

Semi-discontinuous Replication: Continuous replication of one strand and discontinuous replication of another strand of DNA.

Sense Strand: The DNA strand which is used as template for synthesis of RNA.

Template: A DNA strand used for replication.

Unidirectional Replication: Replication of DNA, from the point of origin, in one direction only.

QUESTIONS

1. **What is DNA replication? Describe briefly main points related to DNA replication.**
2. **What are the models of DNA replication? Give brief account of any one model.**
3. **Explain briefly the differences in DNA replication in prokaryotes and**

eukaryotes.

4. **Describe briefly the various steps involved in semi-conservative replication of DNA replication.**

5. **Describe briefly the steps involved in Cairns model of circular DNA replication.**

6. **Explain briefly rolling circle model circular DNA replication.**

7. **What do the following terms signify?**
   (i) Leading strand (ii) Lagging strand
   (iii) Unidirectional replication (iv) Bidirectional replication

8. **Write short notes on the following:**
   (i) Continuous replication (ii) Discontinuous replication
   (iii) Okazaki Fragments (iv) Origin

9. **Describe the role of different enzymes in DNA replication in prokaryotes and eukaryotes.**

10. **How would you prove that DNA replicates in semi-conservative manner?**

11. **Explain the following:**
    (i) Meselson and Stahl experiment sopporting semi-conservative replication of DNA.
    (ii) Cairns experiment sopporting semi-conservative replication of DNA.
    (iii) Taylor's experiment sopporting semi-conservative replication of DNA.

*****

Chapter 4

RNA Structure and Types

INTRODUCTION

Ribonucleic acid (RNA) is a nucleic acid polymer consisting of nucleotide monomers, that act as a messenger between DNA and ribosomes and is also responsible for making proteins out of amino acids. It carries the genetic message from DNA to ribosomes and is involved in the process of protein synthesis. Ribonucleic acid is one of the two types of nucleic acids found in all cells. The other is deoxyribonucleic acid (DNA). RNA is similar to DNA but containing ribose in place of deoxyribose and uracil in place of thymine. Some viruses use RNA instead of DNA as their genetic material. The main features of RNA are presented as follows:

In eukaryotes, DNA is found both in nucleus and cytoplasm. In the nucleus it is a component of chromosome, whereas in cytoplasm it is found in ribosomes. In prokaryotes, it is found in the cytoplasm. The RNA synthesis takes place in the nucleus on DNA template. After synthesis it moves from nucleus to cytoplasm. Thus RNA usually does not have self-duplication property. In certain viruses like TMV and plantango virus RNA can synthesize another RNA molecule.

There are different forms of RNA such as (i) messenger RNA [mRNA, (ii) ribosomal RNA [r-RNA], and(iii) transfer RNA [t-RNA] or soluble RNA [s-RNA] Ribosomal and transfer RNAs constitute about 98 per cent of the total RNA. All three forms of RNA are synthesized on DNA template. RNA molecule is much smaller in size than DNA. It consists up to 12,000 nucleotides whereas DNA consists up to 4.3 million nucleotides.

In most of the organisms, the usual function of RNA is transfer of genetic message from nucleus to the cytoplasm and synthesis of protein in ribosomes. In some viruses, RNA acts as the genetic material and regulates the gene action.

STRUCTURE OF RNA

Mostly the RNA molecule is single stranded in both eukaryotes and prokaryotes. However, in some viruses RNA is double stranded. The double stranded RNA is found in Rio virus in animals and wound tumor virus in plants.

The single strand of RNA is folded upon itself either entirely or in some regions. Most of the base in the folded region are complementary and are joined by hydrogen bonds. However, in the unfolded regions the bases do not have complements. As a result, the purine and pyrimidine ratio is not equal in RNA molecule.

The RNA is composed of ribose sugar, nitrogenous bases and phosphate group. The nitrogenous base in RNA are: adenine (A), guanine (G), cytosine (C), and uracil(U). In RNA, uracil is present in place of thymine of DNA molecule. In RNA, the sugar is ribose, while in DNA it is deoxyribose.

COMPONENTS OF RNA

The RNA molecule is a polymer which is composed of several thousand pairs of nucleotide monomers. The polymer is known as a "polynucleotide." Each nucleotide consists of a 5-carbon sugar (ribose), a nitrogen containing base attached to the sugar, and a phosphate group. Thus there are three components of RNA, *viz.* (1) nitrogenous bases, (2) ribose sugar, and (3) phosphate group. These are briefly discussed below:

1. Nitrogenous Bases

Nucleotides are also known as nitrogenous bases or DNA bases. Nitrogenous base are of two types, *viz.* pyrimidines and purines. Pyrimidines are single ring structures. These are of two types namely cytosine and thymine.They occupy less space in DNA structure. Pyrimidine is linked with deoxyribose sugar at position 3.

Purines are double ring compounds. They are of two types, *viz.* adenine and guanine.

They occupy more space in DNA structure. Deoxyribose sugar is linked at position 9 of purine. Thus, in DNA there are four different types of nitrogenous bases, *viz.* adenine (A), guanine (G), cytosine (C) and thymine (T). In RNA, the pyrimidine base thymine is replaced by uracil.

Thus, in RNA there are four different types of nitrogenous bases, *viz.* adenine (A), guanine (G), cytosine (C) and uracil (T). In DNA, the pyrimidine base uracil is replaced by thymine.

Base Pairing

The purine and pyrimidine bases always pair in a definite fashion. Adenine will always pair with uracil and guanine with cytosine. Adenine and thymine are joined by double hydrogen bonds while guanine and cytosine are joined by triple hydrogen bonds.

2. Ribose Sugar

RNA contains ribose sugar while DNA contains deoxyribose sugar. Both are pentose sugar having five carbon atoms. The four carbon atoms are inside the ring and the fifth one is with CH_2 group. This has four OH groups on 1, 2, 3 and 5 carbon positions. Hydrogen atoms are attached to carbon atoms one to four. The ribose sugar has OH group on carbon atom 2, whereas DNA has H group.

3. Phosphate

The phosphate molecule is arranged in an alternate manner to ribose sugar. Thus there is ribose sugar on both sides of phosphate. The phosphate is joined with carbon atom 3 of ribose at one side and with carbon atom 5 of ribose on the other side.

Nucleosides and Nucleotides

In RNA, nucleoside is a combination of ribose sugar and nucleotide is a combination of nucleoside and phosphate. Thus, a nucleotide is a nucleoside with one or more phosphate groups covalently attached to it. Nucleosides differ from nucleotides in that they lack phosphate groups. The four different nucleosides of RNA are ribose adenosine (rA), ribose guanosine (rG), ribose cytosine (rC), and ribose urosine (rU).

RNA Backbone

The RNA backbone is a polymer with an alternating sugar-phosphate sequence. The ribose sugar is joined at both the 3′-hydroxyl and 5′-hydroxyl groups to phosphate groups in ester links, also known as "phosphodiester" bonds.

DIFFERENCES BETWEEN DNA AND RNA

There are some similarities between DNA and RNA and some difference which are presented in the Table 4.1.

TABLE 4.1: Comparison of DNA and RNA Molecules

Sl.No.	*Particulars*	*DNA Deoxyribose Nucleic Acid*	*Ribose Nucleic Acid*
1	Location	Nucleus and cytoplasm	Chromosomes and ribosomes
2	Replication	Self replication	Self replication rare
3	Types	Several	Three: m-RNA, t-RNA, and r-RNA
4	Strands	Usually two, rarely one	Usually one, rarely two
5	Bases	A, G, C and T	A, G, C and U
6	Base pairing	AT and GC	AU and GC
7	Sugar	Deoxyribose	Ribose
8	Size	Up to 4.3 million nucleotides	Up to 12,000 nucleotides
9	Function	Genetic role	Protein synthesis; genetic in some viruses.

TYPES OF RNA

On the basis of function RNA is of two types, *viz.* (1) genetic RNA, and (2) non-genetic RNA. The non-genetic RNA is of three types, *viz.* (i) transfer RNA, (ii) messenger RNA, and (iii) ribosomal RNA. A brief discussion of all these types of RNA is presented as follows:

Genetic RNA

RNA which acts as genetic material like DNA is called genetic RNA. Such RNA is found in most of the plant viruses [TMV,HRV, *etc.*], some animal viruses and certain bacteriophages. Genetic RNA may be single stranded or double stranded. Genetic RNA has self-replication property.

Non-Genetic RNA

RNA which does not act as genetic material is known as non-genetic RNA. This is found in higher organisms where DNA is the genetic material. Such RNA is usually single stranded. This type of RNA does not have self-replication property. Such RNA is synthesized from DNA template in the presence of DNA dependent RNA polymerase enzyme. Thus genetic RNA differs from non-genetic RNA in several aspects.

TABLE 4.2: Comparison of Genetic and Non-genetic RNA Molecules

Sl.No.	*Particulars*	*Genetic RNA*	*Non-genetic RNA*
1	Strand	Either single or double	Always single
2	Types	Only one type	Three types
3	Replication	Self replication	Synthesized from DNA
4	Organisms in which found	Plant and animal viruses and certain bacteriophages	Plants, animals and bacteria
5	Function	Regulation of gene action	Protein synthesis
6	Present	In the absence of DNA	In association with DNA

Messenger RNA

The RNA which carries information from nuclear DNA to cytoplasm for protein synthesis is referred to as messenger RNA. It is also known as template RNA. It constitutes about 5 to 10 per cent of the total cellular DNA. The main features of messenger RNA are presented below:

1. The new mRNA is synthesized during early cleavage on a DNA strand in the presence of RNA polymerase enzyme. In eukaryotes heterogeneous nuclear RNA [hnRNA] is synthesized from DNA. This is considered as precursor of mRNA.The hnRNA may be 5000 to 50,000 thousand nucleotides long. It consists of both coding and non-coding nucleotide sequences. The non-coding sequences are removed during processing. The mature mRNA may be a friction of original hn-RNA. Synthesis of mRNA differs from DNA replication in following three main aspects.
   (i) Ribose nucleotides are used instead of deoxyribose nucleotides.
   (ii) Adenine pairs with uracil instead of thymine.
   (iii) Only one strand of DNA take part in the synthesis of mRNA.
2. Usually each gene transcribes its own RNA. There are as many types of mRNA molecules as there are genes. There may be 1000-10,000 types of mRNA in a cell. These types of mRNA differ only in the sequence of their bases and in their length.
3. The molecular weight of an average molecule of mRNA is 500,000 and its sedimentation coefficient is 8 S.

4. In bacteria, it is short lived. For example, in *E. coli* the average half life of some mRNA is about 2 minutes. However, in mammals it may live for many hours and even days.
5. The main function of mRNA is protein synthesis in ribosomes in the cytoplasm.

Ribosomal RNA

RNA found in ribosomes in the cytoplasm. is called ribosomal RNA. It constitutes about 80 per cent of the total cellular RNA. Its main features are given below:

1. In eukaryotes, ribosomal RNA is synthesized from nuclear DNA. In prokaryotes it is synthesized from a part of DNA. Synthesis of ribosomal RNA begins during gastrulation and increases as embryo develops.
2. On the basis of molecular weight and sedimentation rate, ribosomal RNA is of following three types:
   (i) Molecular weight over a million [2S-29S RNA]
   (ii) Molecular weight below one million [12 S-18S], and
   (iii) Low molecular weight [5S RNA]
3. Ribosomal RNA is more stable than mRNA.
4. The function of ribosomal RNA is binding of mRNA and tRNA to ribosomes in the cytoplasm.

Transfer RNA

RNA which carries amino acids and attach them with ribosome mRNA complex for use in protein synthesis is termed as transfer RNA. It is also known as soluble RNA.It constitutes about 10 to 15 per cent of the total RNA of the cell. Main features of transfer RNA are given below:

1. Transfer RNA is synthesized on a DNA template using small section of the DNA molecule. It is synthesized at the end of cleavage.
2. There are more than a hundred types of t-RNA per bacterial cell. There are 20 amino acids, hence there should be at least 20 types of t-RNA. However, t-RNA is always more than 20 and in some cases, there are two types of t-RNA which specify for one amino acid.
3. It has molecular weight of about 25,000 to 30,000 with sedimentation coefficient of 3.8 S. It contains 73 to 93 nucleotides.
4. The main function of t-RNA is to carry various types of amino acids and attach them to mRNA template for protein synthesis.

SUMMARY

A nucleic acid that carries the genetic message from DNA to ribosomes and is involved in the process of protein synthesis is referred to as RNA [ribose nucleic acid].Ribonucleic acid is one of the two types of nucleic acids found in all cells.

Mostly the RNA molecule is single stranded in both eukaryotes and prokaryotes. However, in some viruses RNA is double stranded. The double stranded RNA is found in Rio virus in animals and wound tumor virus in plants.

The RNA molecule is a polymer which is composed of several thousand pairs of nucleotide monomers. There are three components of RNA, *viz.* (1) nitrogenous bases, (2) ribose sugar, and (3) phosphate group.

The RNA is of two types, *viz.* (1) genetic RNA, and (2) non-genetic RNA. The non-genetic RNA is of three types, *viz.* (i) transfer RNA, (ii) messenger RNA, and (iii) ribosomal RNA.

GLOSSARY

Base Pairing: The pairing of purine and pyrimidine bases in DNA and RNA molecules.

Double Stranded RNA: RNA having double strands and is found in some viruses such as Rio virus in animals and tumor producing virus in plants.

Genetic RNA: RNA which acts as genetic material. Such RNA is found in most of the plant viruses [TMV, HRV *etc.*], some animal viruses and certain bacteriophages.

Messenger RNA: The RNA which carries information from nuclear DNA to cytoplasm for protein synthesis.

Monocistronic RNA: The mRNA which is coded by one cistron [gene].

Non-genetic RNA: RNA which does not act as genetic material. It is found in higher organisms where DNA is the genetic material.

Nucleoside: A combination of deoxyribose sugar and nitrogen base.

Nucleotide: A combination of deoxyribose sugar, nitrogen base and phosphate linkages.

Polycistronic RNA: The mRNA which is coded by several cistrons [genes].

Purines: Nitrogenous base with double ring structure.

Pyrimidines: Nitrogenous base with single ring structure.

Ribose Sugar: The sugar which is similar to deoxyribose except that it has OH group on carbon atom 2 instead of H group.

Ribosomal RNA: RNA which is found in ribosomes in the cytoplasm.

RNA: A nucleic acid that carries the genetic message from DNA to ribosomes and is involved in the process of protein synthesis.

Single Stranded RNA: RNA having one helix only; a usual form of RNA.

Transfer RNA: RNA which carries amino acids and attach them with ribosome mRNA complex for use in protein synthesis; also known as soluble RNA.

QUESTIONS

1. **What is RNA? Describe its important features.**
2. **What are different components of RNA molecule? Describe them briefly.**
3. **What are the differences between DNA and RNA?**
4. **What are different types of RNA? Describe any one of them in detail.**
5. **Compare genetic RNA and non-genetic RNA in tabular form.**
6. **What are the difference between eukaryotic mRNA and prokaryotic mRNA?**
7. **Write short notes on the following:**

   (i) Ribose Sugar (ii) Messenger RNA

   (iii) Transfer RNA (iv) Ribosomal RNA
8. **Describe briefly the functions of RNA.**
9. **How would you prove that in certain viruses RNA is the genetic material?**
10. **What do the following terms signify?**

    (i) Genetic RNA (ii) Non-genetic RNA

    (iii) Monocistronic RNA (iv) Polycistronic RNA

*****

Chapter 5

DNA Markers

INTRODUCTION

Marker characters refer to those traits that can be easily identified. DNA marker refers to any DNA sequence having a known location on a chromosome and associated with a particular gene or trait. In other words, any unique DNA sequence which can be used in DNA hybridization, PCR or restriction mapping experiments to identify that sequence is called DNA marker. Thus DNA markers also known as genetic markers refer to variations in DNA fragments generated by restriction endonuclease enzymes.

TYPES OF MARKERS

Markers are of four types, *viz.* (i) morphological, (ii) biochemical, (iii) cytological, and (iv) DNA markers. These are briefly discussed as follows:

1. Morphological

In plant breeding, markers that are related to variation in shape, size, color and surface of various plant parts are called morphological markers. Such markers refer to available gene loci that have obvious impact on morphology of plant. Genes that affect form, coloration, male sterility or resistance among others have been analyzed in many plant species. Morphological markers have been identified in almost every field crop. There are several demerits of morphological markers as given below:

1. They generally express late into the development of an organism. Hence their detection is dependent on the development stage of the organism.
2. They usually exhibit dominance.
3. Sometimes they exhibit deleterious effects.
4. They exhibit pleiotropy.
5. They exhibit epistasis.

6. They exhibit less polymorphism.
7. They are highly influenced by the environmental factors.

2. Biochemical

Markers that are related to variation in proteins and amino acid banding pattern are known as biochemical markers. A gene encodes a protein that can be extracted and observed; for example, isozymes and storage proteins.

3. Cytological

Markers that are related to variation in chromosome number, shape, size and banding pattern are referred to as cytological markers. In other words, it refers to the chromosomal banding produced by different stains; for example, G banding.

4. DNA Markers

DNA markers are also known as molecular markers or genetic markers. DNA marker refers to any DNA sequence, occurring in proximity to the gene or locus of interest. It refers to any unique DNA sequence which can be used in DNA hybridization, PCR or restriction mapping experiments to identify that sequence. It can be identified by a range of molecular techniques such as RFLPs, RAPDs, AFLP, SNPs, SCARs, microsatellites, *etc.* To overcome problems associated with morphological markers, the DNA-based markers have been developed. Advantages of DNA markers are presented below:

1. They are highly polymorphic.
2. They have simple inheritance (often co-dominant).
3. They abundantly occur throughout the genome.
4. They are easy and fast to detect.
5. They exhibit minimum pleiotropic effect.
6. Their detection is not dependent on the developmental stage of the organism.

TABLE 5.1: Comparison of Morphological and DNA Markers

Sl.No.	*Particulars*	*Morphological Markers*	*DNA Markers*
1	Nature	Dominant	Co-dominant
2	Polymorphism	Low	High
3	Occurrence	Low	Abundant
4	Detection	Easy	Easy
5	Pleiotropic effect	High	Minimum
6	Epistasis	Present	Absent
7	Environmental effect	High	No effect

PROPERTIES OF AN IDEAL DNA MARKERS

An ideal DNA marker should have some properties or characteristics. Important properties of an ideal DNA marker are presented below:

1. Markers should exhibit high level of polymorphism. In other words, there should be variability in the markers. It should demonstrate measurable differences in expression between trait types and/or gene of interest.
2. Marker should be co-dominant. It means, there should be absence of intra-locus interaction. It helps in identification of heterozygotes from homozygotes.
3. The marker should be multi-alleleic. It is useful in getting more variability/ polymorphism for a character.
4. There should be absence of epistasis. It makes identification of all phenotypes (homo- and heterozygotes) easy.
5. The marker should be neutral. The substitution of alleles at the marker locus should not alter the phenotype of an individual. This property is found in almost all the DNA markers.
6. Markers should be insensitive to environment. This property is also found in almost all the DNA markers.

TYPES OF DNA MARKERS

There are several types of DNA markers that are used in genetics and plant breeding for various purposes. The use of a specific marker technique is determined by the crop species, technical expertise, lab equipment, and research funding. DNA markers could be broadly categorized into two types:

Hybridization Based

1. Restriction Fragment Length Polymorphism (RFLP),
2. Diversity Array Technology (DArT),
3. Variable Number of Tandem Repeats (VNTR)

PCR Based

1. Randomly Amplified Polymorphic DNA (RAPD)
2. Amplified Fragment Length Polymorphism (AFLP)
3. Simple Sequence Repeat (SSR) Length Polymorphism
4. Inter Simple Sequence Repeats (ISSR)
5. Sequence Characterized Amplified Region (SCAR)
6. Cleaved Amplified Polymorphic Sequences (CAPS)
7. Single Nucleotide Polymorphism (SNP)

A brief description of some of the important mentioned DNA markers is presented as follows:

1. Restriction Fragment Length Polymorphisms (RFLPs)

RFLPs refer to variations found within a species in the length of DNA fragments generated by specific endonuclease. RFLPs are first type of DNA markers developed to distinguish individuals at the DNA level. RFLP technique was developed before the discovery of Polymerase Chain Reaction [PCR].

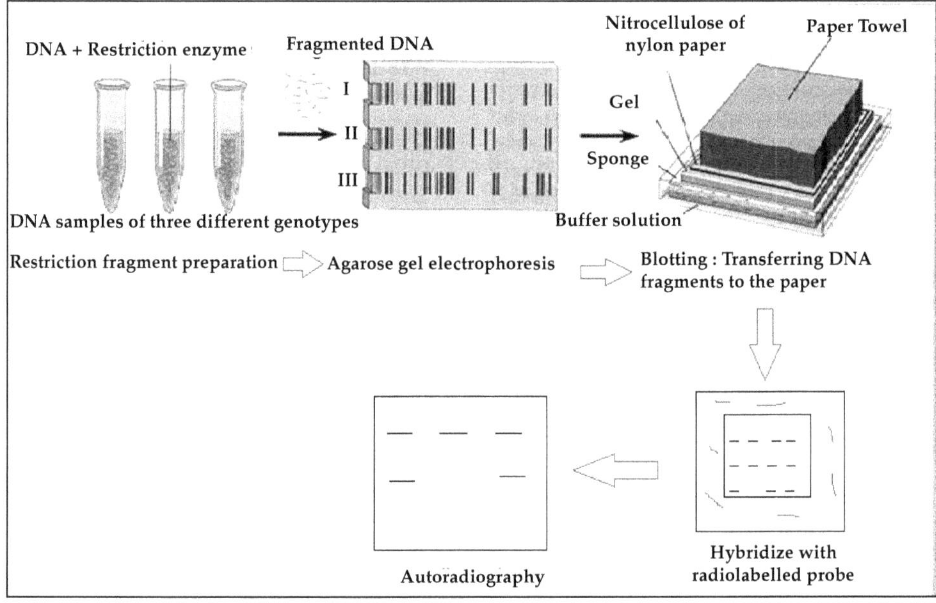

FIGURE 5.1: Steps in RFLP Production.

Steps for RFLP

- Genomic DNA isolation and incubation with Restriction Enzymes
- DNA fragments separated via electrophoresis and transfer to nylon or nitrocellulose membrane
- Membranes exposed to probes labelled with P^{32} via southern hybridization
- Film exposed to X-Ray

The advantages, disadvantages and uses of this technique are presented below.

Advantages

RFLP technique has several advantages. It is a cheaper and simple technique of DNA sequencing. It does not require special instrumentation. The majority of RFLP markers are co-dominant and highly locus specific. These are powerful tools for comparative and synteny mapping. It is useful in developing other markers such as CAPS and INDEL. Several samples can be screened simultaneously by this technique using different probes. RFLP genotypes for single copy or low copy number genes can be easily scored and interpreted.

Disadvantages

Developing sets of RFLP probes and markers is labor intensive. This technique requires large amount of high quality DNA. The multiplex ratio is low, typically one per gel. The genotyping throughput is low. It involves use of radioactive chemicals. RFLP finger prints for multi-gene families are often complex and difficult to score. RFLP probes cannot be shared between laboratories.

Uses

They can be used in determining paternity cases. In criminal cases, they can be used in determining source of DNA sample. They can be used to determine the disease status of an individual. They are useful in gene mapping, germplasm characterization and marker assisted selection. They are useful in detection of pathogen in plants even if it is in latent stage.

2. Amplified Fragment Length Polymorphism (AFLP)

AFLP assays are performed by selectively amplifying a pool of restriction fragments using PCR. RFLP technique was originally known as selective restriction fragment amplification.

Steps Involved in AFLP

- ❒ Extraction of DNA by CTAB or kit method
- ❒ Restriction digestion using hexa-cutter and tetra cutter
- ❒ PCR using primers specific to adapter sequence
- ❒ Second amplification by PCR
- ❒ PAGE or capillary electrophoresis to resolve the fragments

The advantages, disadvantages and uses of this technique are presented below:

Advantages

It has very high multiplex ratio and genotyping throughput. Results are highly reproducible across laboratories. No marker development work is needed; however, AFLP primer screening is often necessary to identify optimal primer specificities and combinations. No special instrumentation is needed for performing AFLP assays; however, special instrumentation is needed for co-dominant scoring. Start-up costs are moderately low. AFLP assays can be performed using very small DNA samples (0.2 to 2.5 µg per individual). The technology can be applied to virtually any organism with minimal initial cost.

Disadvantages

The maximum polymorphic information content for any bi-allelic marker is 0.5. High quality DNA is needed to ensure complete restriction enzyme digestion. DNA quality may or may not be a weakness depending on the species. Developing locus specific markers from individual fragments can be difficult and does not seem to be widely done. The switch to non-radioactive assays has not been rapid. The fingerprints produced by fluorescent AFLP assay methods are often difficult to interpret and score and thus do not seem to be widely used. AFLP markers often densely cluster in centromeric regions in species with large genomes, *e.g.*, barley (*Hordeum vulgare* L.) and sunflower (*Helianthus annuus* L.).

Uses

This technique has been widely used in the construction of genetic maps containing high densities of DNA marker. In plant breeding and genetics, AFLP markers are used in varietal identification, germplasm characterization, gene tagging and marker assisted selection.

3. Simple Sequence Repeats (SSRs)

Simple sequence repeats (SSRs) or microsatellites are tandemly repeated mono-, di-, tri-, tetra-, penta-, and hexa-nucleotide motifs. SSR length polymorphisms are caused by differences in the number of repeats. SSR loci are individually amplified by PCR using pairs of oligonucleotide primers specific to unique DNA sequences flanking the SSR sequence. The advantages, disadvantages and uses of this technique are presented below:

Advantages

SSR markers tend to be highly polymorphic. The genotyping throughput is high. This is a simple PCR assay. Many SSR markers are multi-allelic and highly polymorphic. SSR markers can be multiplexed, either functionally by pooling independent PCR products or by true multiplex-PCR. Semi-automated SSR genotyping methods have been developed. Most SSRs are co-dominant and locus specific. No special equipment is needed for performing SSRs assays. However, special equipment is needed for some assay methods, *e.g.*, semi-automated fluorescent assays performed on a DNA Sequencer. Start-up costs are low for manual assay methods (once the markers are developed). SSR assays can be performed using very small DNA samples (~100 ng per individual). SSR markers are easily shared between laboratories.

Disadvantages

The development of SSRs is labor intensive. SSR marker development costs are very high. SSR markers are taxa specific. The start-up cost is high for automated SSR assay methods. Developing PCR multiplexes is difficult and expensive. Some markers may not multiplex.

Uses

SSR markers are used for mapping of genes in eukaryotes. The most widely-used markers in major cereals are SSRs or 'microsatellites. SSRs are PCR-based markers and extremely versatile, since they are used for basic and applied research. They are highly reliable (*i.e.*, reproducible), co-dominant in inheritance, highly polymorphic (compared to other markers) and generally transferable between mapping populations.

4. Single Nucleotide Polymorphism (SNP)

The variations which are found at a single nucleotide position are known as single nucleotide polymorphisms or SNP. Such variation results due to substitution, deletion or insertion. This type of polymorphism has two alleles and also called bi-alleleic loci. This is the most common class of DNA polymorphism. It is found both in natural lines and after induced mutagenesis.

The advantages, disadvantages and uses of this technique are presented below:

Advantages

SNP markers are highly polymorphic and mostly bi-allelic. The genotyping throughput is very high. SNP markers are locus specific. Such variation results due to substitution, deletion or insertion. SNP markers are excellent long term investment. SNP markers can be used to pinpoint functional polymorphism. This technique requires small amount of DNA

Disadvantages

Most of the SNPs are bi-allelic and less informative than SSRs. Multiplexing is not possible for all loci. Some SNP assay techniques are costly. Development of SNP markers is labor oriented. More (three times) SNPs are required in preparing genetic maps than SSR markers.

Uses

SNPs are useful in preparing genetic maps. They have been used in preparing human genetic maps. In plant breeding, SNPs have been used to lesser extent. SNP markers are useful in gene mapping. SNPs help in detection of mutations at molecular level. SNP markers are useful in positional cloning of a mutant locus. SNP markers are useful in detection of disease causing genes.

4. Cleaved Amplified Polymorphic Sequences (CAPS)

CAPS polymorphisms are differences in restriction fragment lengths caused by SNPs or INDELs that create or abolish restriction endonuclease recognition sites in PCR amplicons produced by locus-specific oligonucleotide primers.

(i) CAPS assays are performed by digesting locus-specific PCR amplicons with one or more restriction enzymes and separating the digested DNA on agarose or polyacrylamide gels.

(ii) CAPS analysis is versatile and can be combined with single strand conformational polymorphim (SSCP), sequence-characterized amplified region (SCAR), or random amplified polymorphic DNA (RAPD) analysis to increase the chance of finding a DNA polymorphism.

(iii) Michaels and Amasino (1998) proposed a variant of the CAPS method called CAPS based on SNPs.

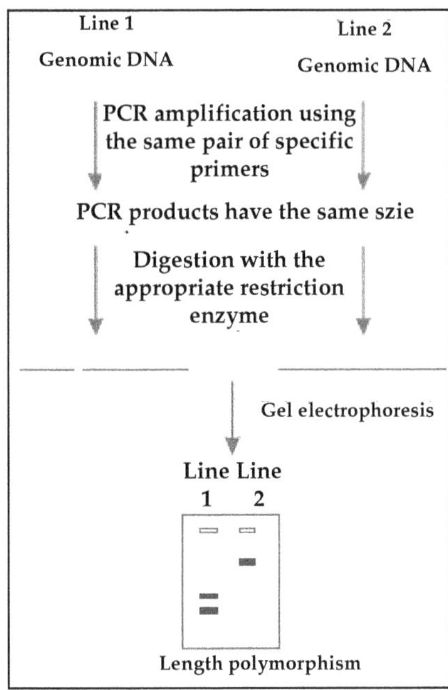

FIGURE 5.2: Steps in CAPS Technique.

The advantages, disadvantages and uses of this technique are presented below.

Advantages

The genotyping throughput is moderately high. It is a simple PCR assay. Markers are developed from the DNA sequences of previously mapped RFLP markers. Most CAPS markers are co-dominant and locus specific. No special equipment is needed to perform manual CAPS marker assays. CAPS marker assays can be performed using semi-automated methods, *e.g.*, fluorescent assays on a DNA Sequencer. The start-up costs are low for manual assay methods. CAPS assays can be performed using very small DNA samples (typically 50 to100 n. g. per individual). Most CAPS genotypes are easily scored and interpreted. CAPS markers are easily shared between laboratories.

Disadvantages

Typically, a battery of restriction enzymes must be tested to find polymorphisms. Although CAPS markers still have great utility and should not be overlooked, other methods have emerged as tools for screening locus-specific DNA fragments for polymorphisms, *e.g.*, SNP assays. The development of easily scored and interpreted assays may be difficult for some genes, especially those belonging to multi-gene families.

Uses

This is straightforward way to develop PCR-based markers from the DNA sequences of previously mapped RFLP markers.

TABLE 5.2: Comparison of RFLP, RAPD, SSR, CAPS and SNP Markers

Sl. No.	*Characteristics*	*RFLP*	*RAPD*	*SSR*	*CAPS*	*SNP*
1	Level of polymorphism	Medium	Very high	High	Moderate	High
2	Cost	Expensive	Cheap	Expensive	Cheap	Variable
3	Allelism	Co-dominant	Dominant	Co-dominant	Mostly co-dominant	Co-dominant
4	Time required	Time consuming	Quick working	Quick working	Quick	quick
5	Banding pattern	Locus specific	Multi locus	Locus specific	Locus specific	Locus specific
6	Probe/primer	Probe	Primer	Primer	Primer	Primer
7	DNA required (ng)	10000	40	30-60	30-60	5-10

APPLICATIONS IN CROP IMPROVEMENT

DNA markers have several useful applications in crop improvement. The important applications are listed as follows:

1. DNA markers are useful in the assessment of genetic diversity in germplasm, cultivars and advanced breeding material.
2. DNA markers can be used for constructing genetic linkage maps, *i.e.*, mapping of genes and quantitative trait loci (QTLs).
3. DNA markers are useful in identification of new useful alleles in the germplasm and wild species of crop plants.
4. DNA markers are used in the marker assisted or marker aided selection. MAS has several advantages over straight selection.
5. DNA markers are useful in the study of crop evolution.
6. Prediction of heterotic cross combinations
7. Gene Pyramiding
8. Positional cloning
9. Selection of parents for hybridization

SUMMARY

Those characters which can be easily identified are called marker characters. Markers related to variations in DNA fragments generated by restriction endonuclease enzymes are called DNA markers or genetic markers. Markers are of four types, *viz.* (i) morphological, (ii) biochemical, (iii) cytological, and (iv) DNA markers. All these have been discussed. Properties of ideal DNA markers have been listed.

There are several types of DNA markers that are used in genetics and plant breeding for various purposes. The commonly used DNA markers include: Restriction Fragment Length Polymorphism (RFLP), Amplified Fragment Length Polymorphism (AFLP), Simple Sequence Repeat (SSR) Length Polymorphism, Single Nucleotide Polymorphism (SNP) and Cleaved Amplified Polymorphic Sequences (CAPS). Among them SSR is most commonly used markers in crop improvement programe and most recent molecular marker is SNP. DNA markers have several useful applications in crop improvement. Important applications include: assessment of diversity, gene mapping, discovery of new genes, marker assisted selection and study of crop evolution.

GLOSSARY

AFLP: Amplified Fragments Length Polymorphism.

Biochemical markers: The protein markers.

CAPS: Cleaved Amplified Polymorphic Sequences are differences in restriction fragment lengths caused by SNPs or INDELs.

Cytological markers: The chromosomal markers.

DNA fingerprinting: A technique by which an individual can be identified on the basis of DNA sequences.

DNA marker technology: A technology that utilizes molecular markers for genetic improvement of crop plants such as marker assisted selection; also called molecular marker technology.

DNA marker: See molecular marker.

Marker assisted selection: Indirect selection for a desired plant phenotype based on the banding pattern of linked molecular marker.

Molecular back crossing: Back crossing based on molecular markers.

Molecular breeding: A branch of plant breeding which utilizes molecular genetic tools and approaches for genetic improvement of crop plants.

Molecular genotyping: Molecular characterization of breeding material.

Molecular marker technology: A technology that utilizes molecular markers for genetic improvement of crop plants such as marker assisted selection.

Molecular marker: A unique sequence of nucleotides found on a strand of DNA.In other words, a distinctive stretch of DNA located close to the gene of interest.

Molecular selection: Same as marker assisted selection.

Morphological markers: Markers related to morphology such as colour and shape.

QTL mapping: A technique of locating genes which control expression of quantitative traits.

Quantitative trait: A character showing scontinuous variation and governed by polygenes.

RFLP: Restriction Fragment Length Polymorphism, a molecular DNA marker technique.

SNP: Single Nucleotide Polymorphism, a type of molecular marker used in genetic engineering.

SSR: Simple Sequence Repeats, a type of DNA marker.

QUESTIONS

1. **What is a marker? List commonly used DNA markers and describe any one of them in detail.**
2. **Describe merits and demerits of morphological markers?**
3. **What are the advantages of molecular markers over morphological markers?**
4. **Describe important features of an ideal DNA marker.**
5. **Give a brief comparison of morphological markers and DNA markers.**
6. **Give a comparison of Restriction Fragment Length Polymorphism (RFLP) and Amplified Fragment Length Polymorphism (AFLP).**
7. **What are the applications of DNA markers in crop improvement?**
8. **Define RFLP and AFLP markers and discuss their advantages, disadvantages and uses.**
9. **Define RAPD and SSR markers and discuss their merits, demerits and uses.**
10. **Define CAPS markers and discuss their advantages, disadvantages and uses.**
11. **Define SNP markers and discuss their advantages, disadvantages and uses.**

*****

Chapter 6

Marker Assisted Selection

INTRODUCTION

Marker characters refer to those traits that can be easily identified. Marker assisted selection is also known as marker aided selection (MAS). It is a process whereby a marker (morphological, biochemical or one based on DNA/RNA) is used for indirect selection of a specific trait. Marker assisted selection is based on genetic markers. Marker assisted selection (MAS) is indirect selection process where a trait of interest is selected on a marker linked to it. MAS can be used as an aid to traditional phenotypic-pedigree-based selection systems. MAS can be useful for traits that are difficult to measure, exhibit low heritability, and/or are expressed late in development.

GENE VERSUS MARKER

The gene of interest is directly related with production of protein(s) that produce certain phenotypes whereas markers should not influence the trait of interest but are genetically linked (remain together during segregation). In many traits genes are discovered and can be directly assayed for their presence with a high level of confidence. However, if a gene is not isolated marker's help is taken to tag a gene of interest. In such case there may be some false positive results due to recombination between marker of interest and gene (or QTL). A perfect marker would elicit no false positive results.

SELECTION OF MAJOR GENE LINKED TO MARKERS

The major genes control oligogenic or monogenic characters which are economically important. Such characteristics include disease resistance, male sterility, self-incompatibility, others related to shape, color, and surface of plants are often of mono- or oligogenic in nature. The marker loci which are tightly linked to major genes can be used for selection and are sometimes more efficient than direct selection for the target gene. Such advantages in efficiency may be due, for example, to higher expression of the marker mRNA in such cases that the marker is actually a gene. Alternatively, in such cases that the target gene of interest differs between two alleles by a difficult-to-detect single nucleotide polymorphism, an external marker (be it another gene or a polymorphism that is easier to detect, such as a short tandem repeat) may present as the most realistic option.

SITUATIONS FOR USE OF MAS

There are several situations where the use of molecular markers in the selection of a genetic trait would be rewarding. Such situations include (i) late expression of the trait, (ii) trait controlled by recessive gene, (iii) resistance to biotic stresses, and (iv) presence of epistasis. These are briefly discussed below:

1. **Late expression of a Trait**: MAS will be useful when the trait of interest is expressed late in plant development, like flower and fruit features.
2. **Control by Recessive Gene**: MAS will be useful when the character of interest is is governed by recessive gene. It can identify both dominant and recessive genes.
3. **Resistance to Biotic Stress**: MAS will be useful in screening the material for resistance to diseases and insects especially in those areas field inoculation with the pathogen is not allowed for safety reasons. Moreover, it eliminates undesirable environmental effects.
4. **Epistasis**: MAS will be useful when the phenotype is affected by two or more unlinked genes (epistatis). For example, selection for multiple genes which provides resistance against diseases or insect pests for gene pyramiding.

The cost of genotyping (an example of a molecular marker assay) is reducing while the cost of phenotyping is increasing particularly in developed countries thus increasing the attractiveness of MAS as the development of the technology continues.

STEPS IN MAS

The marker assisted selection consists of two important steps, *viz.* (i) mapping of gene, and (ii) use of this information for marker assisted selection. These are discussed as follows:

(i) Gene Mapping

The first step is to map the gene or quantitative trait locus (QTL) of interest. This is done by using different techniques. Each gene mapping technique has some merits and demerits. Detailed discussion of all such techniques is beyond the scope of this discussion. Five types of populations, *viz.* (i) recombinant inbred lilen(RILs), (ii) near isogenic lines (NILs), (iii) back cross, (iv) double haploids (DH), and (v) F_2 segregations are commonly used for gene mapping or gene tagging. Linkage between the phenotype and markers which have already been mapped is tested in these populations in order to determine the position of the QTL. Such techniques are based on linkage and are therefore referred to as "linkage mapping". This consists of following steps:

1. Selection of parents with divergent alleles or contrasting traits.
2. Development of mapping populations.
3. Isolation of DNA.
4. Scoring of DNA markers such as RFLPsor AFLPs, *etc.*

(ii) Marker Assisted Selection

The second important step is to use above information for marker assisted selection. Generally, the markers to be used should be close to gene of interest (<5 recombination unit or cM) in order to ensure that only minor fraction of the selected individuals will be recombinants. Generally, two markers are used in order to reduce the chances of an error due to homologous recombination. For example, if two flanking markers are used at same time with an interval between them of approximately 20cM, there is higher probability (99 per cent) for recovery of the target gene. This consists of following steps:

1. Indirect selection using molecular markers, and
2. Correlation of DNA markers with morphological markers.

Various type of DNA markers are used in marker assisted selection. The commonly used markers include RFLPs, RAPDs, AFLP, SNPs, SCARs, microsatellites, *etc.*

APPLICATIONS OF MARKERS

In most major crops, various linked markers exist to screen key traits. Numerous markers have been mapped to different chromosomes in several crops including rice, wheat, maize, soybean and several others. These markers have been used in diversity analysis, parentage detection, DNA fingerprinting, and prediction of hybrid performance. Molecular markers are useful in indirect selection processes, enabling manual selection of individuals for further propagation. In addition to these, applications of MAS in backcrossing and gene pyramiding are discussed below:

1. Marker Assisted Backcross

A minimum of five or six-backcross generations are required to transfer a gene of interest from a donor (may not be adapted) to a recipient (recurrent – adapted cultivar). The recovery of the recurrent genotype can be accelerated with the use of molecular markers. If the F_1 is heterozygous for the marker locus, individuals with the recurrent parent allele(s) at the marker locus in first or subsequent backcross generations will also carry a chromosome tagged by the marker.

2. Marker Assisted Gene Pyramiding

Gene pyramiding has been proposed and applied to enhance resistance to disease and insects by selecting for two or more than two genes at a time. For example in rice such pyramids have been developed against bacterial blight and blast. The advantage of use of markers in this case allows to select for QTL-allele-linked markers that have same phenotypic effect.

ADVANTAGES OF MAS

Marker-assisted selection is the most widely used application of DNA markers. Once traits have been mapped and a closely linked marker has been found, it is possible to screen large numbers of samples for rapid identification of progeny that carry desirable characteristics. MAS has several advantages such as (i) high speed, (ii) consistency of results, (iii) bio-safety, (iv) high efficiency, and (v) QTL mapping. These are briefly discussed below:

1. **Speed:** It is a rapid method of selection of desirable plants. DNA can be extracted from tissue from the first leaves or the cotyledons of a plant. Trait information can be discovered with markers prior to pollination allowing more informed crosses to be made.
2. **Consistency:** The results of MAS are consistent, because use of markers eliminates the impact of environmental variation that often complicates phenotypic evaluation.
3. **Biosafety:** MAS is a safe method of screening. When we use MAS, there is no need of introducing into breeding populations. Particularly for livestock breeding this delivers a very important level of bio-safety.
4. **Efficiency:** MAS has very high efficiency. Screening of progeny in early stages allows a breeder to reject undesirable genotypes from the program more quickly. Most breeding programs that use markers still evaluate the same number of plants in the field however the level of genetic quality is vastly increased because of the early-stage screening that has been carried.
5. **Complex traits:** DNA markers permit mapping of polygenic traits or quantitative trait loci [QTL] which is not possible through conventional plant breeding techniques.

LIMITATIONS OF MAS

Marker assisted selection has some limitations such as (i) expensive technique, (ii) requires technical skill, (iii) a laborious work, and (iv) may sometimes lead to health hazards. These are briefly discussed below:

1. **Expensive Technique**: MAS is an expensive technique, because this technique requires very costly equipments, glassware and chemicals; sophisticated laboratory.
2. **Technical skill:** This technique requires well trained man power for handling of equipments, isolation of DNA molecules and study of DNA markers.
3. **Laborious Work**: The detection of various DNA markers [RFLP, AFLP, RAPD, SSR, SNP, *etc.*] is a laborious and time consuming work. For such markers huge plant breeding population has to be screened to get meaningful results.
4. **Health Hazards**: Some of the DNA marker techniques involve use of radioactive isotopes in labeling of DNA, which may lead to serious health hazards. Now non-radioactive labelling agents are also available.

SUMMARY

Marker assisted selection or marker aided selection (MAS) is a process whereby a marker (morphological, biochemical or one based on DNA/RNA) is used for indirect selection of a specific trait. Markers are of four main types, *viz.* (i) morphological, (ii) biochemical, (iii) cytological, and (iv) DNA markers.

There are several situations where the use of molecular markers in the selection of a genetic trait would be rewarding. Such situations include (i) late expression of the trait, (ii) trait controlled by recessive gene, (iii) resistance to biotic stresses, and (iv) presence of epistasis.

The marker assisted selection consists of two important steps, *viz.* (i) mapping of gene, and (ii) use of this information for marker assisted selection.

MAS has been used in several crops such as rice, wheat, maize, soybean and many others. DNA markers have been used in diversity analysis, parentage detection, DNA fingerprinting, and prediction of hybrid performance. Molecular markers are useful in indirect selection processes, enabling manual selection of individuals for further propagation. Besides these, MAS is used in backcrossing and gene pyramiding.

MAS has several advantages such as (i) high speed, (ii) consistency of results, (iii) bio-safety, (iv) high efficiency, and (v) QTL mapping.

Marker assisted selection has some limitations such as (i) expensive technique, (ii) requires technical skill, (iii) a laborious work, and (iv) may sometimes lead to health hazards.

GLOSSARY

Biochemical Marker: Markers that are related to variation in proteins and amino acid banding pattern.

Cytological Marker: Markers that are related to variation in chromosome number, shape and size.

DNA Marker: A gene or other fragment of DNA whose location in the genome is known. A genetic marker is a known DNA sequence.

Epistasis: The interaction between genes. Epistasis takes place when the action of one gene is modified by one or several other genes.

Genotyping: The process of determining the genotype of an individual with a biological assay

Locus (plural loci): A fixed position on a chromosome, such as the position of a gene or a biomarker (genetic marker).

Marker assisted selection: Selection for specific alleles (which affect a trait of interest) using genetic markers.

Marker: Any genetic element (locus, allele, DNA sequence or chromosome feature) which can be readily detected by phenotype, cytological or molecular techniques, and used to follow a chromosome or chromosomal segment during genetic analysis.

Molecular Breeding: Improvement of crop plants for various economic characters through indirect selection for linked molecular markers.

Morphological Marker: In plant breeding, markers that are related to variation in shape, size, color and surface of various plant parts.

Phenotype: Any observable characteristic of an organism, such as its morphology, development, biochemical or physiological properties, or behavior.

Pleiotropy: Influence of a single gene on multiple phenotypic traits.

Polymorphism: Presence of multiple alleles of a gene within a population, usually expressing different phenotypes.

QUESTIONS

1. **Define marker assisted selection and explain briefly its advantages.**
2. **Describe briefly disadvantages of marker assisted selection.**
3. **Discuss briefly various steps involved in marker assisted selection.**
4. **Define DNA marker and give important features of an ideal marker.**
5. **What are the demerits of morphological markers?**
6. **Give a brief comparison of morphological markers and DNA markers.**
7. **Write short notes on the following:**
   (i) Marker assisted selection (ii) Molecular breeding
8. **Describe briefly marker assisted selection in relation to backcrossing and gene pyramiding.**
9. **Discuss briefly various applications of marker assisted selection in crop improvement.**

*****

Chapter 7

Gene Regulation in Prokaryotes

INTRODUCTION

Gene regulation is the process by which the cell determines [through interactions among DNA, RNA, proteins, and other substances] when and where genes will be activated and how much gene product will be produced. Thus, the gene expression is controlled by a complex of numerous regulatory genes and regulatory proteins. The gene regulation has been studied in both prokaryotes and eukaryotes.

BASICS OF GENE REGULATION

The operon refers to a group of closely linked genes which act together and code for various enzymes of a particular biochemical pathway. In other words, operon is a unit of bacterial gene expression and regulation, including structural genes and control elements in DNA recognized by regulator gene product(s). Thus operon is a model which explains the on-off mechanism of protein synthesis in a systematic manner. The main points of operon model of gene regulation are presented below:

The operon model of gene regulation was developed by Jacob and Monod in 1961 for which they were awarded Nobel Prize in 1965. Now this model of gene regulation is widely accepted. The operon model was developed working with lactose region [lac region] of human intestine bacteria *E. coli*. The gene regulation was studied for degradation of the sugar lactose.

GENES INVOLVED

In the operon model of gene regulation, four types of genes, *viz.*, (i) structural genes, (ii) operator gene, (iii) promoter gene, and (iv) regulator gene are involved. In addition, repressor, co-repressor, and inducer molecules are also involved. A brief description of genes involved in operon model of gene regulation is presented as follows:

1. Structural Genes

There are three structural genes of the lac operon, *i.e., lac* Z, *lac* Y and *lac* A. The main function of structural genes is to control protein synthesis through messenger RNA. Function of these genes is as follows:

(i) *lac Z*: It encodes the enzyme beta-galactosidase, which catalyses the break down of lactose into glucose and galactose.

(ii) *lac Y: It* encodes the enzyme galactosidase permease, which permits entry of lactose from the medium into the bacterial cell.

(iii) *lac A:* It encodes the enzyme transacetylase, which transfers an acetyl group from acetyl co-enzyme A to beta galactosidase.

2. Promoter Gene

The above three structural genes genes are under the control of the promoter gene [designated P]. In the promoter RNA polymerase binds to the DNA and prepares to initiate transcription. The main function of promoter gene is to initiate mRNS transcription.

3. Operator Gene

The other regulatory element in an operon is the operator (designated O). This is the element that determines whether or not the genes of the operon are transcribed. The main function of operator gene is to control function of structural genes.

4. Regulator Gene

This is designated as I. It expresses all the time, or constitutively and plays an important role in operon function. This is the *lac I* gene, which encodes a protein called the lac repressor. The lac repressor has two functional domains or regions: one that binds to the DNA of the operator region, and one that binds to lactose. When the repressor binds to the operator, it prevents RNA polymerase advancing along the operon, and transcription does not occur. The regulation of the operon depends on regulating whether or not the repressor binds to the operator. The function of regulator gene is to direct synthesis of repressor, a protein molecule. Its function differs in the presence and absence of lactose as discussed below:

Types of Regulation

In prokaryotes, operons are of two types, *viz.*, inducible and repressible. The example of an inducible operon is the lactose operon, which contains genes that encode enzymes responsible for lactose metabolism. An example of repressible operon is the Trp operon, which encodes enzymes responsible for the synthesis of the amino acid tryptophan (trp for short).

1. Inducible Operon

Inducible Enzyme: An enzyme whose production is enhanced by adding the substrate in the culture medium is called inducible enzyme. And such system is called inducible system. The example of an inducible operon is the **lactose operon,**

which contains genes that encode enzymes responsible for lactose metabolism. In bacteria, operon refers to a group of closely linked genes which act together and code for the various enzymes of a particular biochemical pathway. The model of lac operon of *E. coli* looks like this:

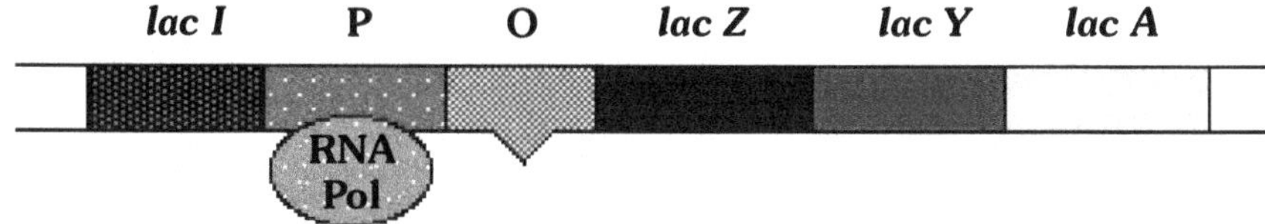

When Lactose is Absent

When the lactose is absent in the environment, events take place in this way. The *lac I* gene is transcribed [constitutively, *i.e.*, continuously] and the mRNA is translated, producing the lac repressor. The repressor binds to the operator, and blocks RNA polymerase. When RNA polymerase is blocked, there is no transcription. Thus the enzymes for lactose metabolism are not synthesized, because there is no lactose to metabolize. Thus when lactose is absent, lactose-metabolizing enzymes are not produced.

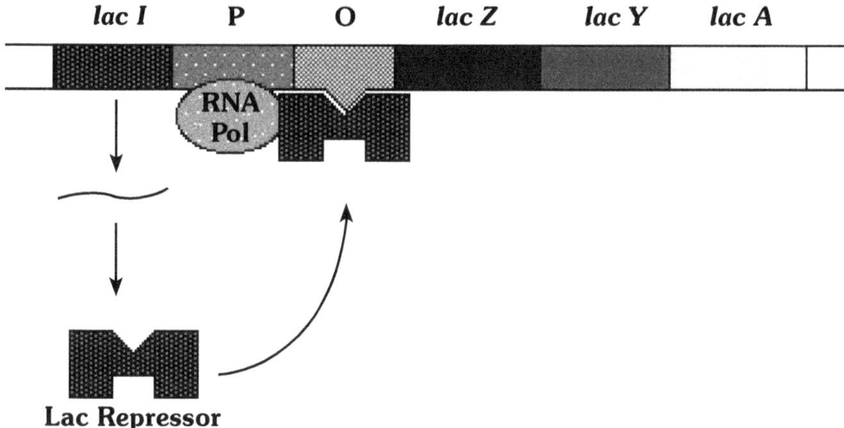

When Lactose is Present

When the lactose is present in the environment, the events occur in a different way. A small amount of the lactose enters the cell, and affects regulation of the operon:

The lac repressor is still synthesized. The repressor can bind to lactose. After binding to lactose, the repressor undergoes a **conformational change** (change of shape). Molecules that change shape when they bind to another molecule are called allosteric molecules. With this change, the lac repressor is unable to bind to the operator region. Hence RNA polymerase is not blocked, and is able to transcribe the genes of the operon. The enzymes encoded by those genes are produced. The lac permease transports more lactose into the cell, and beta-galactosidase cleaves the lactose into glucose and galactose. This can be further metabolized by other enzymes, producing energy for the cell. Lactose, therefore, is able to **induce** the synthesis of the enzymes necessary for its metabolism (by preventing the action of the repressor). As such, lactose is the inducer of the lac operon.

Thus when lactose is absent, lactose-metabolizing enzymes are not produced, and when lactose is present, those enzymes are produced.

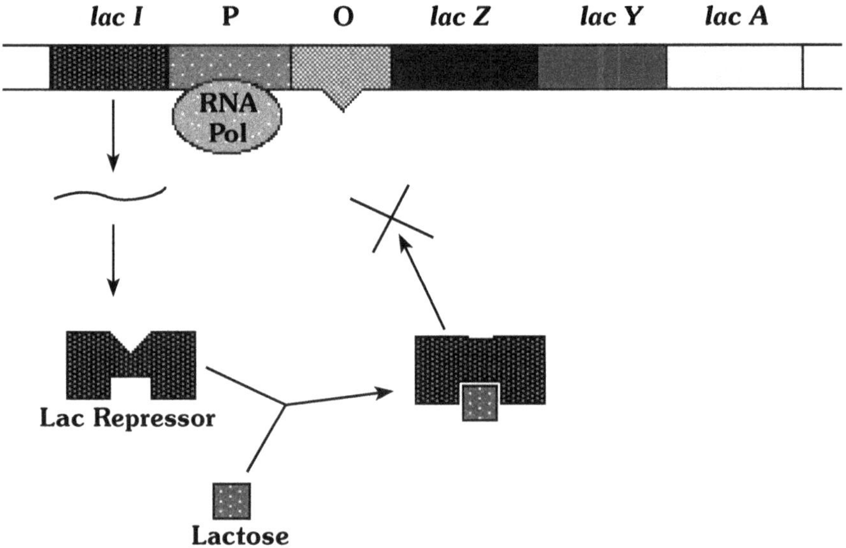

Catabolite Repression

Expression of the *lac* operon can also be regulated in another way. Glucose is preferable to lactose as an energy source. Hence if glucose is present in the environment, the transcription is reduced or *lac* operon is down-regulated.

Transcription of the *lac* operon requires another protein, called catabolite activator protein (CAP for short). This CAP protein binds to the *lac* promoter and enhances transcription. But it occurs only after CAP binds to a small molecule called cyclic AMP (cAMP). Without cAMP, CAP will not bind to the promoter, and no transcription will occur. In the previous examples involving the *lac* operon, we can assume that cAMP was present, and the CAP-cAMP complex was bound to the promoter.

The cAMP is produced by an enzyme called adenylcyclase. In the presence of glucose in the environment, adenylcyclase is inhibited, and cAMP production drops. Thus there is no cAMP to bind to CAP. In this situation, the CAP will not bind to the *lac* promoter, and no *lac* transcription takes place. In this way, the bacterium does not produce enzymes for lactose metabolism when they are not necessary because of the presence of glucose. Beta-galactosidase breaks lactose in to glucose and galactose. When enough lactose has been metabolized, glucose (one of the products) accumulates and causes repression of the *lac* operon.

2. Repressible Operon

A protein molecule which prevents transcription is called repressor and the process of inhibition of transcription is called repression. Repressible operons are regulated by the end product of the metabolic pathway and not by a reactant in the metabolic pathway (such as lactose in lac operon). An example of repressible operon is the **Trp operon.** This encodes enzymes which are responsible for the synthesis of the amino acid **tryptophan** (trp for short). The trp operon is regulated by trp, which is the product of the metabolic pathway.

In trp operon, the **trp repressor** only binds to the operator when trp is **present**. (opposite to the *lac* repressor). The repressor binds to trp, and undergoes a conformational change [change of shape]. This change in shape allows it to bind to the operator, blocking transcription. Because trp is needed for repression, it is referred to as a **co-repressor** in this system (as opposed to lactose being an inducer). When trp is absent, the repressor will not bind to the operator, and transcription occurs. Thus, if there is plenty of trp around [and no more is needed], the transcription is blocked. If there is no trp around [it needs to be synthesized], transcription occurs. In other words, it allows production of the enzymes for trp synthesis.

MECHANISM OF GENE REGULATION

The mechanism of gene regulation is of two types, *viz.* (1) negative regulation, and 2) positive regulation. The mechanism of gene regulation in *E. coli* operon and tryptophan operon are discussed below:

1. Negative Control

The first switch in the *lac* operon of *E. coli*, is the repressor protein. In negative control, the transcription is controlled by repressor protein, which is an allosteric protein. The repressor protein binds to operator region and prevents transcription. It prevents transcription by blocking RNA polymerase. Thus, when repressor is bound to operator, the transcription is switched off. Thus the on-off switch of protein synthesis is governed by free or occupied position of the operator gene. When the operator is free, transcription will take place and when the operator gene is blocked, the transcription is prevented.

If an isomer of lactose [allo-lactose] is present, it will bind to repressor protein and change its shape. The changed repressor does not bind to operator and thus allows transcription.

2. Positive Control

The second switch in the *lac* operon of *E. coli* is the catabolite activator protein [CAP].The CAP is an allosteric protein. The CAP binds to DNA and small molecule called cyclic adenosine mono phosphate [cAMP]. The CAP only binds to promoter region and stimulates transcription when cAMP binds to allosteric site.

The concentration of cAMP is controlled by ATP concentrations. The low ATP leads to high cAMP and high ATP leads to low cAMP. If *E. coli* is growing on glucose, there will be high [ATP] and low [cAMP]. If no glucose is present, there will be a low [ATP] and high [cAMP].

In the absence of glucose, [cAMP] is high, binds to CAP which binds to promoter region and stimulates transcription. If glucose is present, [cAMP] is low, doesn't bind to CAP which can not bind to promoter and doesn't allow transcription.

MERITS OF OPERON MODEL

1. It is a very simple yet informative model of gene regulation in prokaryotes.
2. It is a very well understood model of gene regulation in prokaryotes.
3. This model is based on empirical results and has been studied on different prokaryotes.
4. This model is of two types, *viz.*, (i) inducible operon, and (ii) repressible.

SUMMARY

Gene regulation is the process by which the cell determines [through interactions among DNA, RNA, proteins, and other substances] when and where genes will be activated and how much gene product will be produced. Thus, the gene expression is controlled by a complex of numerous regulatory genes and regulatory proteins.

In prokaryotes, the operon model of gene regulation is widely accepted. This model of gene regulation was proposed by Jacob and Monod in 1961 for which they were awarded Nobel Prize in 1965. The operon is a unit of bacterial gene expression and regulation, including structural genes and control elements in DNA recognized by regulator gene product(s). Thus operon is a model which explains the on-off mechanism of protein synthesis in a systematic manner.

In prokaryotic gene regulation, a group of closely linked genes work together in operon under the control of a single promoter and operator and code for various enzymes of a particular biochemical pathway. Transcription occurs when repressor is not bound to operator. The binding of repressor with operator inactivates the operator and prevents transcription.

In the operon model of gene regulation, four types of genes, *viz.*, (i) structural genes, (ii) operator gene, (iii) promoter gene, and (iv) regulator gene are involved. In addition, repressor, co-repressor, and inducer molecules are also involved.

In prokaryotes, operons are of two types, *viz.*, inducible and repressible. The example of an inducible operon is the lactose operon and example of repressible operon is the Tryptophan operon. In an inducible system, the repressor binds to the operator unless it is bound by an inducer. An inducer is a molecule at the start of the metabolic pathway which is governed by the enzymes encoded by the operon genes. In *lac* operon of *E. coli*, there are two types of control of gene regulation, *viz.*, (i) negative control and (ii) positive control.

In a repressible system, the repressor only binds to the operator if it binds to a co-repressor. The co-repressor is an end product of the metabolic pathway which is governed by the enzymes encoded by the operon genes. The mechanism of gene regulation has been discussed for *E. coli* operon and tryptophan operon.

GLOSSARY

Allosteric Molecules: Molecules that change shape when they bind to another molecule.

Constitutive Transcription: When transcription is carried out continuously.

Co-repressor: A combination of repressor and metabolite which prevents protein synthesis. Such process is termed as co-repression.

Costitutive Enzyme: An enzyme whose production is constant irrespective of of the metabolic state of the cell.

Effector: The molecule which acts as an inducer or co-repressor in the operon model of *E. coli*.

Gene Regulation: The study of on-off mechanism of protein synthesis; also known as regulation of gene action, regulation of gene expression and regulation of protein synthesis.

Inducer: The substance which allows initiation of transcription (lactose in *lac* operon). Such process is known as induction.

Inducible Enzyme: An enzyme whose production is enhanced by adding the substrate in the culture medium. Such system is called inducible system.

Negative Regulation: Inhibition of transcription by repressor through inactivation of promote such as in *lac* operon of *E. coli*.

Operator gene: In the *lac* operon of *E. coli*, a gene which controls the function of structural genes.

Operon Model: A group of closely linked genes which act together and code for the various enzymes of a particular biochemical pathway.

Positive Regulation: Enhancement of transcription by an effector molecule through activation of promoter.

Promoter gene: In the *lac* operon of *E. coli*, a gene which initiates mRNA transcription. It is located between regulator and operator genes.

Regulator gene: In the *lac* operon of *E. coli*, a gene found at the end and directs the synthesis of repressor protein molecule.

Repressible Enzyme: An enzyme whose production can be inhibited by adding an end product. Such system is known as repressible system.

Repressor: In the *lac* operon of *E. coli*, a protein molecule which prevents transcription. The process of inhibition of transcription is called repression.

Structural genes: In the *lac* operon of *E. coli*, the genes which control the synthesis of protein through mRNA.

QUESTIONS

1. **Describe briefly different components of prokaryotic operon model.**
2. **Describe the role of regulator, promoter, operator and structural genes in gene regulation in *E. coli*.**
3. **What are types of operon in prokaryotes? Describe inducible operon in detail.**
4. **Define repressible operon and describe the same in detail.**
5. **Explain briefly the mechanism of gene regulation in *E. coli*.**
6. **Write short notes on the following:**
   (i) Structural genes (ii) Operator gene
   (iii) Promoter gene (iv) Regulator gene
7. **Define the following terms:**
   (i) Repressor (ii) Co-repressor
   (iii) Repression (iv) Co-repression
8. **Define the following terms:**
   (i) Inducer (ii) Inducible enzyme
   (iii) Repressible enzyme (iv) Constitutive enzyme
9. **Differentiate between the following:**
   (i) Repressor and inducer
   (ii) Negative and positive gene regulation
   (iii) Inducible and repressible enzymes.
   (iv) Promoter and operator genes
10. **Give a brief account of negative and positive control of gene regulation in *E. coli*.**

*****

Chapter 8

Gene Regulation in Eukaryotes

INTRODUCTION

Gene regulation is the process by which the cell determines [through interactions among DNA, RNA, proteins, and other substances] when and where genes will be activated and how much gene product will be produced. Thus, the gene expression is controlled by a complex of numerous regulatory genes and regulatory proteins. The gene regulation has been studied in both prokaryotes and eukaryotes. This chapter deals with gene regulation in eukaryotes.

BASICS OF GENE REGULATION

In eukaryotes, many models of gene regulation were proposed. However, the model given by R.J. Britten and E.H. Davidson in 1969 became the most popular and is widely accepted. This model is also known as gene battery model.This model is only a theoretical model and lacks sound practical proof. Even then this model is widely accepted for gene regulation in eukaryotes. In the Britten and Davidson model of eukaryotic gene regulation, one enzyme, *viz.* mRNA polymerase is involved in transcription.

GENES INVOLVED

In the Britten and Davidson model of eukaryotic gene regulation, four types of genes, *viz.* (i) sensor gene, (ii) receptor gene, (iii) integrator gene, and (iv) producer gene are involved. A brief description of these genes in relation to gene regulation is presented below:

1. Producer Gene

It is comparable to a structural gene in prokaryotes. It produces pre-mRNA, which after processing becomes mRNA. Its expression is under the control of many receptor sites. Each producer gene may have several receptor sites, each responding to one activator. Thus, though a single activator can recognize several genes, different activators may activate the same gene at different times. This gene is located just near the receptor gene. This gene controls the transcription of mRNA fro DNA and also synthesis of specific proteins. The producer gene starts transcription after receiving signal from receptor gene.

2. Receptor Site (Gene)

It is comparable to the operator in bacterial operon. At least one such receptor site is assumed to be present adjacent to each producer gene.It provides a link between integrator gene and producer gene The activator RNA binds with receptor gene The receptor gene activates the producer gene as soon as it receives signal from the integrator gene through activator RNA. A specific receptor site is activated when a specific activator RNA or an activator protein, a product of integrator gene, complexes with it.

It is also proposed that receptor sites and integrator genes may be repeated a number of times so as to control the activity of a large number of genes in the same cell. Repetition of receptor ensures that the same activator recognizes all of them and in this way several enzymes of one metabolic pathway are simultaneously synthesized.

3. Integrator Gene

Integrator gene is comparable to regulator gene and is responsible for the synthesis of an activator RNA molecule that may not give rise to proteins before it activates the receptor site. At least one integrator gene is present adjacent to each sensor site. It is located very near to the sensor genes. The main function of integrator gene is to start transcription after receiving signal from sensor gene. The RNA which is transcribed by the integrator gene is known as activator RNA. This activator RNA is a link between integrator and receptor genes. Britten and Davidson proposed that the integrator gene products are activator RNAs that interact directly with the receptor genes to trigger the transcription of the continuous producer genes. Transcription of the same gene may be needed in different developmental stages. This is achieved by the multiplicity of receptor sites and integrator genes.

4. Sensor Gene (Site)

These are signal receiving genes. These genes are very sensitive for specific signals from the cell and its environment. Sensor genes are activated by various cellular substances like enzymes, hormones and metabolites. Whenever these genes receive signals, they pass on the message to the adjacent gene (integrator) for initiation of transcription. A sensor site regulates activity of an integrator gene which can be transcribed only when the sensor site is activated. The sensor sites are also regulatory sequences that are recognized by external stimuli, *e.g.*, hormones, temperature.

According to the Britten Davidson model, specific sensor genes represent sequence-specific binding sites (similar to CAP–cAMP binding site in the *E. coil*) that respond to a specific signal. When sensor genes receive the appropriate signals, they activate the transcription of the adjacent integrator genes. The integrator gene products will then interact in a sequence specific manner with receptor genes.

A set of structural genes controlled by one sensor site is termed as a battery. Sometimes when major changes are needed, it is necessary to activate several sets of genes. If one sensor site is associated with several integrators, it may cause transcription of all integrators simultaneously thus causing transcription of several producer genes through receptor sites.

The repetition of integrator genes and receptor sites is consistent with the reports that state that sufficient repeated DNA occurs in the eukaryotic cells. The most attractive features of the Britten and Davidson model is that it provides a plausible reason for the observed pattern of interspersion of moderately repetitive DNA sequences and single copy DNA sequences.

Direct evidence indicates that most structural genes are indeed single copy DNA sequences. The adjacent moderately repetitive DNA sequences would contain the various kinds of regulator genes (sensor, integrator and receptor genes).

MECHANISM OF GENE REGULATION

According Britten and Davidson model, first the sensor gene receives signal from the cell. Then integrator gene becomes active and transcribe activator mRNA molecule. This activator RMA molecule gets connected with receptor gene. Then the receptor gene activates the producer gene, which initiates transcription of mRNA and protein synthesis. In this model, the sensor and integrator genes lie together, whereas receptor and producer genes make another group.

MERITS AND DEMERITS

Merits

1. This model is widely accepted for gene regulation in eukaryotes.
2. The mRNA is stable in eukaryotes which is unstable in prokaryotes.

Demerits

1. This is only a theoretical model and lacks sound practical proof.
2. The mechanism of this model is more complex than operon model of gene regulation in prokaryotes.
3. This model is lesser understood than operon model of gene regulation in prokaryotes.

SUMMARY

Gene regulation is the process by which the cell determines [through interactions among DNA, RNA, proteins, and other substances] when and where genes will be activated and how much gene product will be produced.

In eukaryotes, many models of gene regulation were proposed. However, the model given by R.J. Britten and E.H. Davidson in 1969 became the most popular and is widely accepted. This model is also known as gene battery model. This model is only a theoretical model and lacks sound practical proof. Even then is model is widely accepted for gene regulation in eukaryotes.

In the Britten and Davidson model of eukaryotic gene regulation, four types of genes *viz.*, (i) sensor gene, (ii) receptor gene, (iii) integrator gene, and (iv) producer gene are involved. In the Britten and Davidson model of eukaryotic gene regulation, one enzyme *viz.* mRNA polymerase is involved in transcription.

The main drawbacks of this model are that it is (i) only a theoretical model without proof, (ii) more complicated, and (iii) lesser understood than operon mode.

GLOSSARY

Activator RNA: The RNA which is transcribed by the integrator gene.

Gene Regulation: The study of on-off mechanism of protein synthesis.; also known as regulation of gene action, regulation of gene expression and regulation of protein synthesis.

Integrator gene: In operon model of eukaryotes, a gene located very near to sensor gene. It initiates transcription after receiving signal from sensor gene.

Operon Model: A group of closely linked genes which act together and code for the various enzymes of a particular biochemical pathway.

Producer gene: In operon model of eukaryotes, a gene located very near to receptor gene and is responsible for transcription and protein synthesis.

Receptor gene. In operon model of eukaryotes, a gene located very near to producer gene. It activates producer gene for initiation of transcription.

Sensor genes: Signal receiving genes in eukaryotes operon model proposed by Britten and Davidson in 1969.

QUESTIONS

1. **What are the genes involved in gene regulation in eukaryotes?**
2. **Explain briefly the mechanism of gene regulation in Eukaryotes.**
3. **What do the following terms signify?**
   (i) Sensor genes (ii) Integrator gene
   (iii) Receptor gene (iv) Producer gene
4. **Compare the gene regulation mechanism in prokaryotes and eukaryotes.**
5. **Define sensor, integrator, receptor and producer genes. Describe their role in gene regulation in eukaryotes.**
6. **Differentiate between the following:**
   (i) Sensor and integrator genes
   (ii) Receptor and producer genes

7. **Give a brief account of enzymes involved in eukaryotic gene regulation.**
8. **Describe briefly the battery model of gene regulation in eukaryotes.**
9. **Explain the contribution of the following scientists:**

   (i) Britten and Davidson (ii) Jacob and Monod

*****

Section II

Plant Biotechnology

Chapter 9

History of Plant Biotechnology

INTRODUCTION

Plant biotechnology is a powerful tool for the development of new crop varieties with useful plant traits. Such new plant varieties must be produced on a large scale to achieve commercial success and to satisfy the demand of growers. Plant biotechnology is founded on the principles of cellular totipotency and genetic transformation. Plant biotechnology refers to the development of useful products (varieties or pharmaceuticals) or services from plant cells, tissues or sometimes organs.

IMPORTANT EVENTS

Plant biotechnology is founded on the demonstrated totipotency of plant cells, combined with the delivery, stable integration, and expression of transgenes in plant cells, the regeneration of transformed plants, and the Mendelian transmission of transgenes to the progeny. Important events in the history of tissue culture and plant biotechnology are presented in Tables 9.1 and 9.2.

TABLE 9.1: Plant Tissue Culture

Year	*Name of Scientist(s)*	*Main Contribution*
1902	Haberlandt	Made first attempt of plant tissue culture (Father of Plant Tissue culture).
1904	Hannig	Made first attempt to culture embryo of selected crucifers.
1922	Knudson	Reported asymbiotic germination of orchid seeds *in vitro.*
1922	Robbins	First reported *In- vitro* culture of root tips.
1925	Laibach	First applied embryo culture technique in inter specific hybrids of Linseed (*Linum).*
1934	Gautheret	Tried *In- vitro* culture of the cambial tissue of a few trees and shrubs, although failed to sustain cell division.
1934	White	First reported successful culture of tomato roots.

Year	*Name of Scientist(s)*	*Main Contribution*
1939	Gautheret, Nobecourt and White	Reported successful establishment of continuously growing callus cultures.
1940	Gautheret	Tried *in-vitro* culture of cambial tissues of *Ulmus* to study adventitious shoot formation.
1941	Van Overbeek	First used coconut milk containing a cell division factor for culture of Datura embryos.
1941	Braun	*Observed in- vitro* culture of crown gal tissues.
1944	Skoog	Reported *in-vitro* adventitious shoot formation in tobacco.
1946	Ball	First Raised whole plants of *Lupinus* and *Tropaeolum* by shoot tip culture.
1950	Ball	Reported regeneration of organs from callus tissue of *Sequoia sempervirens.*
1952	Morel and Martin	First used meristem culture to obtain virus-free Dahlias.
1952	Morel and Martin	Reported first application of micro-grafting.
1953	Tulecke	Reported production of haploid callus of the gymnosperm *Ginkgo biloba* from pollen.
1954	Muir *et al.*	Regenerated first plant from a single cell.
1955	Miller *et al.*	First discovered kinetin, a cell division hormone.
1957	Skoog and Miller	Discovered the regulation of organ formation by changing the ratio of auxin : cytokinin.
1958	Maheshwari and Rangaswamy	Regenerated somatic embryos *in-vitro* from the nucellus of Citrus ovules.
1959	Reinert and Steward	First reported the production of embryos from callus and cell suspensions of carrots.
1959	Gautheret	Publication of first handbook on "Plant Tissue Culture".
1960	Kanta	First reported successful test tube fertilization in *Papaver rhoeas.*
1960	E. Cocking	Demonstrated the possibility of enzymatic isolation of protoplast from higher plants.
1960	Bergmann	Filtration of cell suspensions and isolation of single cells by plating.
1960	Barski *et al.*	First used the term somatic cell hybridization.
1962	Murashiqe and Skoog	First proposed a nutrition medium for tobacco tissues regeneration.
1963	Yamada *et al.*	First obtained haploid callus from anther cultures of *Tradescantia reflexa.*
1964	Guha and Maheshwari	First produced haploid plants from pollen grains of Datura (Anther culture).
1967	Bourgin and Nitsch	First obtained mature flowering plants from pollen culture in *N. tobaccum* and *N. sylvestris.*
1968	Takebe *et al.*	First used commercial enzyme preparation for isolation of protoplasts.
1970	Carlson	Selection of biochemical mutants *in vitro* by the use of tissue culture derived variation.
1970	Power *et al.*	Reported first successful achievement of protoplast fusion.
1970	Toshiba Murashige	Developed the concept of synthetic seeds and first used the term artificial seed.

Year	*Name of Scientist(s)*	*Main Contribution*
1971	Heinz and Mee	First reported the morphological variation in plants regenerated from cell culture in sugarcane.
1971	Takebe *et al.*	First regenerated plants from protoplasts.
1972	Carlson *et al.*,	First reported interspecific hybridization through protoplast fusion in two species of *Nicotiana.*
1974	Reinhard	Reported transformation in plant tissue cultures.
1974	Bunn *et al.*	First used the term cybrid.
1976	Seibert	First reported shoot initiation from cryo-preserved shoot apices of carnation.
1976	Power *et al.*	Reported inter-specific hybridization by protoplast fusion in *Petunia hydrida* and *P. parodii.*
1978	Melchers *et al.*	Made somatic hybridization between tomato and potato resulting in pomato, a useless combination.
1979	Marton *et al.*	Developed co-cultivation procedure for transformation of plant protoplasts with *Agrobacterium.*
1980	Hanstein	First used the term protoplast.
1980	Alfermann *et al.*	Use of immobilized whole cells for biotransformation of digitoxin into digoxin.
1981	Larkin and Scowcroft	First introduced the term somaclonal variation.
1981	Gordon and Ruddle	First used the term transgenic.
1982	Krens *et al.*	Incorporation of naked DNA by protoplast resulting in the transformation with isolated DNA.
1982	Zimmermann	First reported fusion of protoplasts using electric stimuli.
1983	Pelletier *et al.*	Intergeneric cytoplasmic hybridization in Radish and Grape.
1986	Powell-Abel *et al.*	Developed TMV virus-resistant tobacco and tomato transgenic plants using cDNA of coat protein gene of TMV.
1988	Redenbaugh *et al.*	Further advanced the concept of artificial seed patented artificial seed technology.
1990	Kranz and coworkers,	First reported successful *in-vitro* fertilization in maize.
2005	A Team of Scientists	Sequenced Rice genome under International Rice Genome Sequencing Project.

TABLE 9.2: Plant Biotechnology

Year	*Name of Scientist(s)*	*Main Contribution*
1960	Kanta	Reported first successful test tube fertilization in *Papaver rhoeas.*
1960	W. Arber	Postulated presence of restriction enzymes.
1964	Hederberg and Meselson	Coined the term Restriction endonuclease.
1968	H.G. Khorana, *et al.*	Deciphered the genetic code.
1968	Meselson and Yuan	Coined the term "Restriction endonuclease" to describe a class of enzymes involved in cleaving DNA.
1970	H. Temin and D. Baltimore	Discovered the presence of reverse transcriptase.

Year	*Name of Scientist(s)*	*Main Contribution*
1970	Smith, Nathan and Arber	Discovered the first restriction endonuclease from *Haemophillus influenza* Rd. It was later purified and named *Hind* II.
1971	Nathans	Prepared first restriction map using *Hind* II enzyme to cut circular DNA or SV 40 into 11 specific fragments.
1972	Berg *et al,*	Produced first recombinant DNA molecule using restriction enzymes.
1974	Zaenen *et al.*; Larebeke *et al.*	Discovered the fact that the Ti plasmid was the tumor inducing principle of *Agrobacterium.*
1977	Maxam and Gilbert	Developed a method of gene sequencing based on degradation of DNA chain.
1978	Smith, Nathan and Arber	Received Nobel prize for the discovery of endonuclease enzymes.
1979	Marton *et al.*	Developed co-cultivation procedure for transformation of plant protoplasts with *Agrobacterium.*
1980	Hanstein	First used the term protoplast.
1981	Gordon and Ruddle	First used the term transgenic.
1982	Krens *et al.*	Reported incorporation of naked DNA by protoplast resulting in the transformation with isolated DNA.
1983	Kary B. Mullis	Discovered Polymerase chain reaction (PCR), a chemical DNA amplification process.
1984	De Block *et al.*; Horsch *et al.*	Developed transgenic tobacco through *Agrobacterium* mediated method of foreign transfer.
1984	Alec Jeffrey	Developed DNA fingerprinting technique for identification of individuals based on DNA sequences.
1986	Powell-Abel *et al.*	Developed TMV virus-resistant tobacco and tomato transgenic plants using cDNA of coat protein gene of TMV.
1987	Sanford *et al.*and Klein *et al.*	Developed biolistic method of foreign gene transfer for plant transformation.
1987	Barton *et al.*	Isolated *Bt.* gene from bacterium (*Bacillus thuringiensis).*
1990	Williams *et al.*; Welsh and McClelland	Developed the Random Amplified Polymorphic DNA (RAPD) technique.
1990	Mariani and co workers	First reported the barnse and barstar system of male sterility in rapeseed.
1991	A Team of Scientists	Developed DNA microarray system using light directed chemical synthesis system.
1995	Vose *et al.*	Developed DNA fingerprinting by Amplified Fragment Length Polymorphism (AFLP) technique.
1997	Blattner *et al.*	Sequencing of E. *coli* genome.
2000	Ingo Potrykus and Peter Beyer	They first developed the Golden Rice. The former was from Institute of Plant Sciences at the Swiss Federal Institute of Technology and the coworker from the University of Freiburg.
2005	Syngenta	A new variety called *Golden Rice 2* was announced which produces up to 23 times more beta-carotene than the original variety of golden rice.

Year	*Name of Scientist(s)*	*Main Contribution*
2005	A Team of Scientists	Sequenced Rice genome under International Rice Genome Sequencing Project
2006	Tuskan, *et al.*	Sequenced genome of Poplur (*Populus tricocrpa*).
2007	Jaillon, *et al.*	Sequenced genome of Grapes (*Vitis vinifera*).
2007	Jaillon, *et al.*	Sequenced genome of Papaya (*Carica papaya*).
2008	Ming, *et al.*,	Sequenced genome of Apple (*Malus domesticus*).
2008	Sato, *et al.*	Sequenced genome of Lotus (*Lotus japonicas*).
2009	Schnable, P *et al.*	Sequenced genome of Corn (*Zea mays*).
2009	Paterson, A *et al.*	Sequenced genome of Sorghum (*Sorghum bicolor*).
2009	Haung *et al.*	Sequenced genome of Cucumber (*Cucumis sativa*).
2010	Schmutz, *et al.*	Sequenced genome of Soybean (*Glycine max*).
2010	AP Chan, *et al.*,	Sequenced genome of Caster bean *(Ricinus communis).*
2010	Vogel, J *et al.*,	Sequenced genome of Brachypodium.
2010	Reccardo Velasco, *et al.*,	Sequenced genome of Peaches (*Prunus persica).*
2010	Eman K Al-Dous *et al.*,	Sequenced genome of Date palm *(Phoenix dactylifera).*
2011	Varshney, el al,	Sequenced genome of Pigeon pea (*Cajanus cajan*).
2011	PGS Consortium,	Sequenced genome of Potato (*Solanum tuberosum*).
2011	Harm van Bakel, *et al.*	Sequenced genome of Cannabis (*Cannabis sativa*).
2011	Bernd Weisshaar, *et al.*,	Sequenced genome of Sugar beet (*Beeta vulgaris*).
2011	Tina T. Hu *et al.,*	Sequenced genome of Arabidopsis lyrata.
2011	Wang, *et al.*	Sequenced genome of Mustard (*Brassica rapa*).
2012	Zheng G *et al.*	Sequenced genome of Foxtail Millet *(Setaria italic).*
2012	Tomato Genome Consortium,	Sequenced genome of Tomato (*Solanum lycopersicum*).
2012	Angelique D'Hont *et al.*	Sequenced genome of Banana (*Musa acuminate*).
2012	Garci- Mas J *et al.,*	Sequenced genome of Melon (*Cucumis melo*).
2012	Wang, *et al.*	Sequenced genome of Flax (*Linum usitatissimum).*
2012	Kunbo Wang *et al.*	Sequenced genome of Cotton (*Gossypium raimondii*).
2012	Angelique D' Hont *et al.*	Sequenced genome of Sweet Orange (*Citrus sinensis*).
2012	Wang and Jhou,	Sequenced genome of Clementine Orange (*Citrus clementina*).
2013	Ling *et al.*	Sequenced genome of *Triticum urartu.*
2013	-	Sequenced genome of *Nicotiana sylvestris L.*
2013	-	Sequenced genome of *Phaseolus vulgaris.*
2014	-	Sequenced genome of *Capsicum annuum.*
2015	CSIR Scientists	Sequenced of Tulsi [*Oscimum sanctum*]
2016	-	Sequenced genome of grass *Diacanthelium.*
2016	-	Sequenced genome of wild species of peanut [*Arachis ipaensis*]
2017	-	Sequenced genome of *Chenopodium quinoa.*

APPLICATIONS OF PLANT BIOTECHNOLOGY

There are three major areas or aspects of plant biotechnology in relation to crop breeding, *i.e.*, Plant tissue culture, transgenic approaches and molecular breeding methods. Thus, applications of plant biotechnology can broadly be classified into three major categories, *i.e.*, (a) tissue culture, (b) transgenics, and (c) marker assisted breeding. Tissue culture and marker assisted selections are non-transgenic applications of plant biotechnology. Applications of plant biotechnology in these three areas are briefly presented as follows: Details of applications are given under respective chapter.

A. Tissue Culture Techniques

Main applications of tissue culture include: Embryo Rescue, Production of Doubled Haploids, Somatic Embryogenesis, Somaclonal Variation, Protoplast Fusion and Cell Suspension Culture.

B. Transgenic Technology

Main applications of transgenic technology include, Herbicide tolerance, Insect and disease resistance, Longer Storage/shelf life and Increased Nutritional Value.

C. Marker Assisted Selection

Important applications of marker assisted selection include:

1. Characterization of Germplasm, Diversity Analyses, Gene Introgression, Gene Pyramiding, Gene Stacking, Combined MAS, Purity Testing, Synteny Study, Backcross Breeding, QTL Mapping, Allele Mining, DNA Barcoding of Plants, and Plant DNA Fingerprinting.

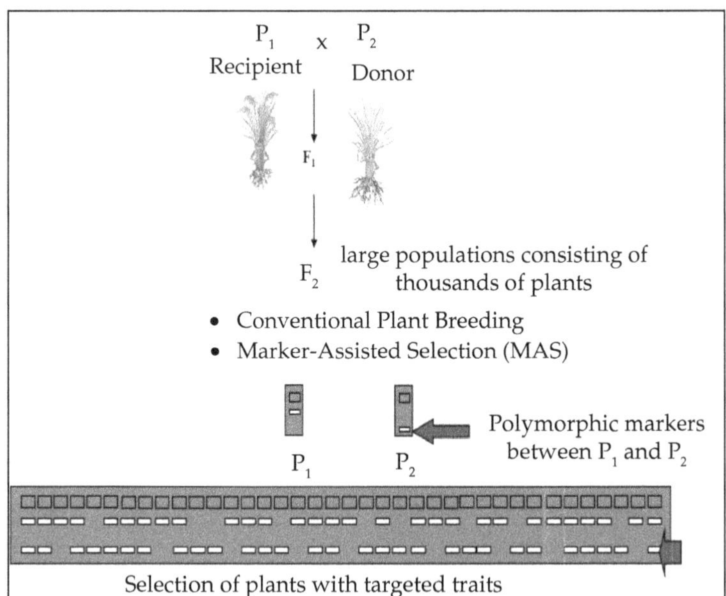

FIGURE 9.1: Selection of Parents with Marker Trait.

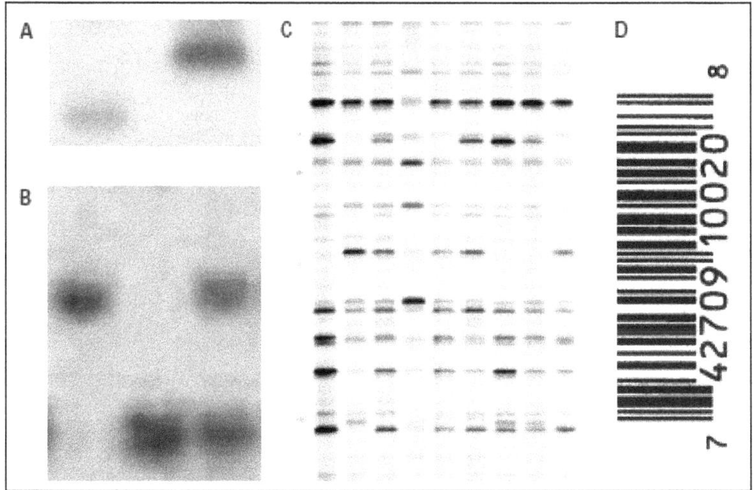

FIGURE 9.2: A, B: Protein Markers, C: DNA Marker, D: Barcoding DNA.

SUMMARY

Plant biotechnology is founded on the principles of cellular totipotency and genetic transformation. Plant biotechnology refers to the development of useful products (varieties or pharmaceuticals) or services from plant cells, tissues or sometimes organs. Important events in the history of molecular biology, tissue culture and plant biotechnology are presented in tabular form.

There are three major areas or aspects of plant biotechnology in relation to crop breeding, *i.e.*, Plant tissue culture, transgenic approaches and molecular breeding methods. Thus, applications of plant biotechnology can broadly be classified into three major categories, *i.e.*, (a) tissue culture, (b) transgenics, and (c) marker assisted breeding. Tissue culture and marker assisted selections are non-transgenic applications of plant biotechnology.

Important applications of tissue culture in crop improvement include embryo rescue, production of doubled haploids, somatic embryogenesis, somaclonal variation, proptoplast fusion and cell suspension culture. These are amply discussed.

In plant breeding, Marker Assisted Selection has several useful applications such as germplasm characterization, diversity analysis, gene introgression, gene pyramiding, gene stacking, combined MAS, purity testing, synteny study, back cross breeding, QTL mapping, allele mining, DNA barcoding, DNA fingerprinting, *etc.* These are briefly discussed.

In plant breeding, Marker Assisted Selection has several useful applications such as germplasm characterization, diversity analysis, gene introgression, gene pyramiding, gene stacking, combined MAS, purity testing, synteny study, back cross breeding, QTL mapping, allele mining, DNA barcoding, DNA fingerprinting, *etc.* These are briefly discussed.

GLOSSARY

Biotechnology: Cellular and bio-molecular processes to develop technologies and products.

Callus: An unorganized mass of differentiated plant cells.

Clonal propagation: A sexual multiplication of plants from a single individual or explant.

Culture: A plant growing *in vitro* in a sterile environment.

Doubled Haploid: Plants derived from a haploid after doubling chromosome number colchicine treatment to form homozygous diploids.

Embryo Rescue: *In vitro* culture technique used to rescue inherently weak, immature or hybrid embryos to prevent degeneration.

***Ex-vitro*:** Organisms removed from tissue culture and transplanted; generally plants to soil or potting mixture.

Explant: An excised piece or part of a plant used to initiate a tissue culture.

Genome editing: Process of making specific changes to the DNA of a cell or organism.

***In vitro*:** To be grown in glass.

In vivo: To be grown naturally.

Marker assisted Selection: A method of selecting desirable individuals in a breeding scheme based on DNA molecular marker patterns.

Micropropagation: Multiplication of plants from vegetative parts by using tissue culture nutrient medium.

Nanotechnology: Study and use of structures between 1 nanometer and 100 nanometers in size.

Protoplast: Cells without cell wall, usually removed by digestion with enzymes.

Somaclonal Variation: Genetic variation among progeny of plants regenerated from somatic cells cultured *in vitro*.

Somatic embryos: Non-zygotic bipolar embryo-like structures obtained from somatic cells.

Tissue Culture: Growth of tissues or cells in an artificial medium.

Totipotency: Capacity of plant cells to regenerate whole plants when cultured on media.

Transgenic: Plants that have a piece of foreign DNA.

QUESTIONS

1. **Define Plant Biotechnology and describe briefly various applications of tissue culture in crop Improvement.**

2. **Explain briefly various applications of transgenic technology in crop improvement.**

3. **Describe role of marker assisted selection and tissue culture techniques in crop improvement.**

4. **Give a brief account of important events in history of plant biotechnology.**

5. **Explain briefly contribution of following scientists in agricultural biotechnology.**

   (i) Griffith (1928)
   (ii) Haberlandt (1902)
   (iii) Toshiba Murashige
   (iv) Guha and Maheshwari (1964)
   (v) Melchers *et al.* (1978)
   (vi) Hanstein (1980)

6. **What is genome editing? How does it differ from transgenic techniques?**

7. **What are advantages and limitations of marker assisted selection?**

8. **What do the following terms signify?**

   (i) Tissue Culture
   (ii) Explant
   (iii) Totipotency
   (iv) Micro-propagation
   (v) Embryo Rescue
   (vi) Doubled Haploids

9. **Write short note on the following:**

   (i) Somaclonal variation
   (ii) Transgenics
   (iii) Nanotechnology
   (iv) Synthetic seed
   (v) Somatic hybridization
   (vi) Callus

10. **Describe in brief various applications of transgenic technology in crop Improvement.**

11. **Write short note on the following:**

    (i) Genome editing
    (ii) Marker assisted selection
    (iii) Protoplast
    (iv) *In vitro* fertilization

*****

Chapter 10

Basics of Plant Tissue Culture

INTRODUCTION

Plant tissue culture is an *in vitro* culture of cells, tissues or organs of plant in a defined nutritional media and aseptic condition to produce an entire plant. German botanist Gottlieb Haberlandt is regarded as the father of tissue culture. He performed his experiment in 1902 on the culture of single cell obtained from cambial tissue of *Acer pseudo-platanus*. It is an important technique for the production of disease free and higher number of plants that are genetically similar to a parent plant within a short period of time. Two concepts, plasticity and totipotency, are the central processes and basic principle to understand the regeneration and plant tissue culture.

IMPORTANT TERMS

Before dealing with tissue culture, it is essential to define some technical terms such as explant, callus, totipotency, embryogenesis and organogenesis.

Explant

The plant part which is used for regeneration in nutrient medium is called explant. It may a meristematic tissue, root tip tissue or any other tissue having totipotency.

Callus

The basis of tissue culture is to grow large number of cells in a aseptic controlled environment. The cells are obtained from stem, root or other plant parts and are allowed to grow in culture medium containing mineral nutrients, vitamins and hormones to encourage cell division and growth. As a result, the cells in culture will produce an unorganised proliferative mass of cells which is known as callus tissue. Callus is unspecialized, unorganized and a dividing mass of cells, formed by the proliferation of the parent tissue. The cells of a callus are parenchymatous, amorphous and unorganised. A callus is produced when explants (cells) are cultured in an appropriate medium. Callus culture involves the growth of a callus (composed of differentiated and non-differentiated cells), which is the followed by a procedure that induces organ differentiation. For this type of tissue culture, the culture is often sustained on a gel medium, which is composed of agar and a mixture of given macro and micronutrients depending on the type of cells. Main points related to callus are presented as follows:

1. **Types:** The callus is generally of two types, *viz.*, friable callus and compact callus. The friable callus can easily be manipulated for suspension culture. However, compact callus is not suitable for suspension culture. In some plant species like carrot and sandalwood somatic embryos are developed but in several crops such as wheat, rice, barley and tobacco development of both root and shoot takes place from the callus.
2. **Color:** The callus may be either white or colored. In some plant species callus is of white color while in other species it may be of light brown or pink color.
3. **Growth:** The growth of callus differs from species to species. The growth on solid or semisolid medium is usually slower than in liquid culture medium. This is attributed to better cell contact and better uptake of nutrient and oxygen in the liquid culture medium.
4. **Callus Induction:** The formation of callus from the explants can be induced by adding auxin in the nutrient medium. The callus induction is easier in dicotyledons than in monocotyledons.
5. **Material Used:** All types of plant parts such as leaves, stems, buds and roots can be used for callus induction. In most of the species, supply of growth regulators such as auxin is required for callus initiation.
6. **Differentiation:** The callus may be differentiated into somatic buds or somatic embryos. In some species, somatic buds are formed from the callus while in some species somatic embryos are formed.

Totipotency

Recent trends of plant tissue culture include genetic modification of plants, production of homozygous diploid plants through haploid cell culture, somatic hybridization, mutation, *etc.* The success of all these studies depends upon the expression of totipotency. Totipotency is the genetic potential of a plant cell to produce the entire plant. In other words, totipotency is the cell characteristic in which the potential for forming all the cell types in the adult organism is retained. Plant breeders, horticulturists and commercial plant growers are now more interested in plant tissue culture only for the exploitation of totipotent cells in culture according to their desirable requirement. Totipotent cells within a bit of callus tissue can be stored in liquid nitrogen for a long period. Therefore, for germplasm preservation of endangered plant species, totipotency can be utilized successfully.

Embryogenesis

The process of formation of somatic embryos from callus cells in culture medium is called embryogenesis. New plants can be obtained from somatic embryos in culture medium.

Organogenesis

The process of differentiation of root and shoot from somatic embryo in the culture medium is called organogenesis.

BASIC REQUIREMENTS

There are some basic requirements of tissue culture techniques, *viz.*, aseptic conditions, proper temperature, proper culture medium, and sub-culturing. These are briefly discussed as follows:

1. Aseptic Conditions

The tissue culture laboratory should have aseptic or well sterilized conditions. The most important factor in tissue culture technique is the maintenance of aseptic conditions. It means it should be well sterilized against pathogens. A pathogen free environment will help in maintaining good health of the callus, cell or protoplast cultures resulting in recovery of healthy plants from such cultures. The explants and glassware should be properly sterilized before their entry into the tissue culture laboratory.

The nutrient media used for plant tissue culture contain sugar as a carbon source, which attracts a variety of micro-organisms such as bacteria and fungi. These organisms grow much faster than the cultured tissues and produce metabolic substances which are toxic to plant tissues. The media may get contaminated by culture vessels, instruments, explant's transfer and culture room. Therefore, sterilization is absolutely essential to provide and maintain a completely aseptic environment during *in vitro* cultivation of plant cells (or) organs.

2. Control of Temperature

Generally, in the tissue culture laboratory temperature is maintained between 18 and 25 degree Celsius. However, the temperature varies from species to species. For maintaining the desired temperature air conditioning is essential. High temperature adversely affects the growth of the callus.

3. Proper Nutrient Medium

Nutrient or culture media has been developed by various workers for different crop species. The nutrient medium has to be modified as per the requirement of a plant species. The culture media developed by Murashige and Skoog (1962) and Gamberg, *et al.*(1968) are used with some modifications in various crop species.

4. Sub-culturing

Transfer of tissues or callus from old culture medium to fresh culture medium is called sub-culturing. This is essential to maintain good health of the callus or tissues, because after some period, some nutrients are depleted in the culture medium and change of medium becomes essential.

BASIC STEPS INVOLVED

There are some basic steps in tissue culture techniques such as explant, culture medium, sterilization, inoculation of explant, incubation, sub-culturing, hardening and transfer of plantlets to the field. These are briefly discussed as follows:

1. Nutrient Medium

The culture medium consists of Inorganic nutrients, Organic nutrients, plant growth hormones and solidifying/gelling agent. Several nutrients media are available to be used in plant tissue culture including MS medium, White's medium, B_5 medium, *etc.* Culture media are largely responsible for the *in vitro* growth and morphogenesis of plant tissues. The success of the plant tissue culture depends on the choice of the nutrient medium. In fact, the cells of most plant cells can be grown in culture media. The composition of the culture media is primarily dependent on particular species of the plant and type of material used for culture, *i.e.*, cells, tissues, organs, protoplasts.

The optimal pH for most tissue cultures is in the range of 5.0-6.0. The pH generally falls by 0.3-0.5 units after autoclaving. Before sterilization, pH can be adjusted to the required optimal level while preparing the medium. It is usually not necessary to use buffers for the pH maintenance of culture media.

At a pH higher than 7.0 and lower than 4.5, the plant cells stop growing in cultures. If the pH falls during the plant tissue culture, then fresh medium should be prepared. In general, pH above 6.0 gives the medium hard appearance, while pH below 5.0 does not allow gelling of the medium.

2. Selection of Explant

Explants are part of plant used in plant tissue culture to grow complete plant. Always young and healthy parts of plants are selected as explants. Explants can be any excised part of the plant to be used in tissue culture like axillary buds, leaf and stem segments, root tip, shoot tip, anther, ovary and endosperm.

3. Sterilization of Explant

Explants are sterilized by using different types of disinfectants like sodium hypochlorite (2 per cent), bromine water (1-2 per cent), silver nitrate (1 per cent) calcium hypochlorite (9-10 per cent), mercuric chloride (0.1-1 per cent), hydrogen peroxide, *etc.* and different types of antibiotics.

4. Inoculation of Explant

The sterile explant is inoculated on the surface of the solidified nutrient medium under aseptic (free from contamination) condition. Generally laminar air flow is used for this inoculation process.

5. Incubation

Incubation is done at 25±2 °C, with a relative humidity of 50-60 per cent with 16 hours of photoperiod. After defined period of incubation, an unorganized and undifferentiated (no root and shoot) mass of cells called callus is obtained from each explant. In some cases, root and shoot directly develop from the explant after incubation.

6. Sub-culturing

The callus obtained from incubation of explant is transferred into another medium and re-incubated. Plantlet regenerates from the callus with the induction of roots and shoot directly.

7. Hardening or Acclimatization

Hardening or acclimatization is the gradual exposure of the plantlets for adjustment to the environmental conditions.

8. Transfer of Plantlets to the Field

After the hardening process, plantlets are transferred to either the greenhouse or the field conditions outside. These plantlets will grow into independent and healthy plants.

APPLICATIONS IN CROP IMPROVEMENT

Plant tissue culture is used widely in crop improvement. Important applications of plant tissue culture are briefly presented as follows:

1. **Production of Virus Free Plants:** Tissue culture technique such as meristem tip culture can be used to produce virus free plants from virus infested stock. This has been achieved in potato and many species of soft fruit.
2. **Production of Synthetic Seeds:** Tissue culture technique such as somatic embryogenesis is used for production of artificial seeds or synthetic seeds.
3. **Production of Haploids:** Haploid plants are obtained by anther culture, pollen culture and unfertilized ovule culture. From haploid plants, homozygous true breeding can be rapidly obtained by doubling chromosome number through colchicine treatment. These are called doubled haploids.
4. **Development of Transgenic Plants:** Tissue culture is useful in development of transgenic plants. After transformation, regeneration of cells or transformed tissues leads to development of transgenic plants.
5. **Embryo Rescue:** Embryo culture technique is useful in saving immature or abortive embryos obtained from distant or incompatible crosses. Such embryos generally die without tissue culture.
6. **Micro-propagation:** It is widely used in forestry and in floriculture. Micro-propagation can also be used to conserve rare or endangered plant species. Micro-propagation using meristem and shoot culture is also used to produce large numbers of identical individuals.

7. **Somatic Hybridization:** Protoplast fusion technology permits production novel hybrids from crosses between distantly related species.
8. **Selection of Resistant Genotypes**: Tissue culture technique is used for selection of cultures resistant to drought, salinity, metal toxicity, and herbicides.
9. **Conservation of Elite Material:** Tissue culture technique is useful in conserving elite materials such as inter-specific hybrids, male sterile lines and novel breeding lines.
10. **Production of Secondary Metabolites:** Valuable chemical compounds, such as plant-derived secondary metabolites and recombinant proteins used as biopharmaceuticals can be easily produced from large-scale growth of plant cells in liquid culture in bioreactors.

LIMITATIONS

There are some limitations of tissue culture technique. Important limitations are listed as follows:

1. It requires well equipped laboratory.
2. It requires costly chemicals and glassware.
3. It requires well trained and adequate staff.
4. It requires adequate funding.

SUMMARY

Growing of living plant cells, tissues or organ on a suitable nutrient medium outside the tissues of parent plant refers too tissue culture. The tissue culture work is carried out in test tube, dish or flask *i.e.* outside the living organism [plant or animal]. Hence, tissue culture techniques are also known as *in vitro* techniques.

There are some basic requirements of tissue culture techniques, *viz.*, aseptic conditions, controlled temperature, proper culture medium, and sub-culturing. These have been explained

Plant tissue culture is the aseptic method of growing various plant parts [meristems, leaves, roots, *etc.* in the nutrient medium under controlled temperature, humidity and light conditions. In this technique small piece of viable tissues called ex-plant is isolated from parent plants and grown in a defined nutritional medium.

There are five important stages of plant tissue culture techniques, *viz.*, pre-initiation stage, initiation stage, proliferation or regeneration stage, differentiation stage and hardening stage. These have been discussed.

GLOSSARY

Aseptic Conditions: The pathogen free environment.

Callus: A mass of unorganized regenerated cells in a culture medium.

Caulogenesis: The process of formation of shoot.

Embryogenesis: The process of formation of somatic embryos from callus cells in culture medium is called embryogenesis. New plants can be obtained from somatic embryos in culture medium.

Explants: The plant part which is used for regeneration.

Hardening: The process of gradual acclimatization of *in vitro* grown plant to *in-vivo* conditions.

Nutrient Medium: A culture medium which contains important nutrients such as essential micro and macro nutrients, carbohydrate, vitamins and hormones.

Organogenesis: The process of differentiation of root and shoot from somatic embryo in the culture medium is called organogenesis.

Rhizogenesis: The process of formation of roots.

Sub-culturing: Transfer of tissues or callus from old culture medium to fresh culture medium.

Totipotency: The regeneration ability of a plant cell to develop into a whole plant in a nutrient medium.

QUESTIONS

1. **Explain briefly basic requirements of plant tissue culture technique.**
2. **Describe briefly applications of plant tissue culture in crop improvement.**
3. **Describe briefly various steps involved in tissue culture techniques.**
4. **Define the following terms:**
   (i) Plant tissue culture (ii) Callus
   (iii) Nutrient medium (iv) Totipotency
   (v) Explant

5. **Write short notes on the following:**
   (i) Aseptic conditions (ii) Sub-culturing
   (iii) Hardening (iv) Limitations of Tissue culture technique

6. **Describe briefly following terms:**
   (i) Embryogenesis (ii) Organogenesis
   (iii) Types of callus (iv) Somatic embryo.

*****

Chapter 11

Nutrient Medium

INTRODUCTION

Explants are grown on a suitable artificially prepared nutrient medium called culture medium. A culture medium which contains important nutrients such as essential micro and macro nutrients, carbohydrate, vitamins and hormones is called nutrient medium. It is also known as culture medium. The pH of the nutrient medium should be 5.5. The composition of culture medium was first given by Murashige and Skoog (1962). In short it is called MS medium. However, the culture medium differs from species to species. Several modifications of Murashige and Skoog medium have been suggested by various workers. A suitable culture medium has to be developed to meet the requirements of a plant species.

TYPES OF NUTRIENT MEDIUM

There are two types of nutrient medium on which explants are grown. These are solid medium and liquid medium. Both media have some merits and demerits. Moreover, there are several differences between these two types of media [Table 11.1]. These are briefly described as follows:

1. Solid Medium

This is the most widely used nutrient medium in plant tissue culture. The semisolid or solid tissue culture medium is prepared by using a solidifying or a gel making agent. The plant material is placed on the surface of the medium. In this medium, the tissue remains intact. However, the cell multiplication rate is slow in this medium.

Advantages

The solid medium is most widely used in plant tissue culture because of its simplicity and easy handling nature. It provides sufficient aeration to explants without a special device because the plant material is placed on the surface of the medium.

Disadvantages

The cell multiplication rate is slow in this medium, because only a part of the explants is in contact with the surface of the medium. This leads to inequality in growth response of tissues and there may be a nutrient gradient between callus and medium. The solid medium represents a static system which leads to variation in the availability of light to the tissue. Moreover, there may be considerable damage to the tissues during sub-culturing. This is not suitable for conducting experiments which require immersion of tissues in the culture medium.

2. Liquid Medium

Liquid medium is used for suspension cultures and for a wide range of research purposes. All the disadvantage of solid medium can be overcome by use of liquid medium. It does not contain a gelling or solidifying agent. So the plant material is immersed in the medium either partially or completely.

Advantages

Liquid medium overcomes all the disadvantages of solid culture medium.

The tissue is more easily supplied with nutrients. The liquid medium represents a dynamic system which leads to equal availability of light to the tissue. The entire explant is in contact with the surface of the medium which leads to faster growth rate of tissues. There will be less damage to the tissues while sub-culturing. In liquid medium isolation of secondary metabolites is easy. It eliminates nutrient gradient within the medium and at the surface of the cells. The removal of toxic waste products is easy. The change of medium is easy. It is suitable for those plant species whose explants exude phenols from their cut surfaces.

Disadvantages

In liquid culture medium, the explants get submerged in liquid medium hence it requires some special devices for proper aeration. The filter paper bridge may be used to keep the explants raised above the level of the medium. It requires regular aeration of cultures through shaker or agitation. Moreover, it requires frequent sub-culturing.

TABLE 11.1: Difference between Solid and Liquid Culture Medium

Sl.No.	*Particulars*	*Solid Medium*	*Liquid Medium*
1	Gel making agent	Required	Not required
2	Aeration to explants	Sufficient	Insufficient
3	Special device for aeration	Not required	Required
4	The type of medium	Static	Dynamic
5	Tissue damage during sub-culturing	High	Low
6	Cell multiplication rate	Slow	Fast
7	Removal of toxic waste products	Difficult	Easy

Sl.No.	Particulars	Solid Medium	Liquid Medium
8	Isolation of secondary metabolites	Difficult	Easy
9	Change of medium	Difficult	Easy
10	Sub-culturing	Less frequent	Frequent
11	Nutrient gradient	Present	Absent

The nutritional requirements for optimum growth of plant organ, tissue and protoplast varies from species to species even tissues from different parts of plant may have different requirement for satisfactory growth. Hence, no single medium can prove satisfactory for all types of *in-vitro* cultures. Nutrient media suggested by Murashige and Skoog and Gamborg *et al.* are widely used.

COMPOSITION OF MS MEDIUM

Nutrient medium is a mixture of certain chemical compounds. It may be a nutrient rich gel or liquid for growing tissue cultures such as cells, organs or plantlets. The culture medium has to supply all the essential mineral ions required for *in-vitro* growth and morphogenesis of plant tissue. The culture medium consists of five main components, *viz.*, (i) inorganic nutrients [micro and macro], (ii) carbon source, (iii) organic supplements, (iv) growth regulators, and (v) solidifying agents. A brief description of all these components is presented as follows:

1. Inorganic Nutrients

This component includes a variety of mineral elements (salts) that supply the macro and micro-nutrients required for the normal growth of plants.

(i) **Macro-nutrients:** Those elements which are required in concentration of 0.5 ml per liter are referred to as macro nutrients. This group includes six major elements, *viz.*, N.P.K.Ca Mg and S. These are present as salts in the media and are essential for plant cell and tissue growth. Nitrogen is the element which is required in greatest amount. It is most commonly supplied as a mixture of nitrate ions (KNO_3) and ammonium ions ($NH_4 NO_3$). Phosphorous is usually supplied as phosphate ion of ammonium, sodium and potassium salts. Other major elements Ca. Mg, and S, are also required to be incorporated in the medium

(ii) **Micro-nutrients:** Those elements which are required in concentration of less than 0.5 ml per liter are considered as micro nutrients. There are eight micro-nutrients, *viz.*, Mn, Zn, B, Cu, Mo, Fe, and Co. The Iron is generally added as a chelate with EDTA (Ethylene Diamino Tetra Acetic acid). In this form iron is gradually released and utilized by living cells and remains available up to a pH of 8. Each nutrient plays important physiological role in plantlet growth and development. Detailed discussion of the role of macro and micro-nutrients is beyond the scope of present discussion.

2. Carbon Source

Sucrose is the most commonly used carbon energy source in plant tissue culture. Generally 2 to 5 per cent of sucrose is used. During autoclaving of the medium sucrose is converted into glucose and fructose. In the process first glucose is used and then fructose, Glucose supports good growth while fructose is less efficient. Maltose galactose, lactose and mannose are the other sources of carbon which are rarely used.

3. Organic Supplements

Organic supplements include vitamins, amino acids and proteins as discussed below:

(i) **Vitamins:** Plants synthesize vitamins endogenously and these are used as catalysts in various metabolic processes. When plant cells and tissues are grown in *in-vitro* some essential vitamins are synthesized but only in suboptimal quantities. Hence, it is necessary to supplement the medium with required vitamins and amino acids to get best growth of tissue.

The most commonly used vitamins is thiamine (vitamin B). Other vitamins which improve growth of cultured plants are Nicotinic acid, Pantothenic acid Pyridoxin (B_6) Folic acid and Aminobenzoic acid.

(ii) **Amino acids:** Cultured tissues are normally capable of synthesizing amino acids necessary for various metabolic processes. In spite of this the addition of amino acid to the media is important for stimulating cell growth in protoplast cultures and for establishing cell culture. Among the amino acids glycine is most commonly used. Other amino acids which are sources of organic nitrogen culture medium include glutamine, aspargine, arginine and cystine.

(iii) **Other organic supplements:** There are some other supplements which are required in culture medium. These include organic extracts, *e.g.*, protein (casein), hydrolysate, coconut milk, yeast and malt extract, ground banana, orange juice, tomato juice, activated charcoal, *etc.* The addition of activated charcoal to culture media stimulates growth and differentiation in orchids, carrot and tomato. Activated charcoal adsorbs inhibitory compounds and darkening of medium occurs. It also helps in reducing toxicity by removing toxic compounds such as phenols produced in the culture.

(iv) **Antibiotics:** The application of some antibiotics such as Streptomycin or Kanamycin at low concentration is essential to prevent the growth of micro-organisms in the culture medium. These antibiotics effectively control systemic infection and inhibit the growth of cell cultures.

4. Growth Regulators

These include auxins, cytokinins, gibberillins, ABA. The growth differentiation and organogenesis of tissue occur only after addition of one or more of these hormones to the medium.

(i) **Auxins:** They have the property of cell division, cell elongation, elongation of stem, internodes, tropism, apical dominance, abscission and rooting. The commonly used auxins include IAA (Indole 3-Acetic Acid), IBA (Indole 3-Butyric Acid), 2,4-D (Dichloro Phenoxy Acetic Acid), NAA (Naphthylene

Acitic Acid) and NOA (Naphthoxy Acitic Acid). The 2,4-D is used for callus induction whereas the other auxins are used for root induction.

(ii) Cytokinins: These are adenine derivatives which are mainly concerned with cell division, modification of apical dominance, and shoot differentiation in tissue culture. Cytokinins activate RNA synthesis and stimulate protein and the enzymatic activity in certain tissues. The commonly used cytokinins are BAP (6-Benzylamino purine), BA (Benzy adenine), 2ip (Isopentyl adenine), Kinetine (6 – furfur aminopurine), Zeatin (4 – hydroxy 3 methyl trans 2 butinyl aminopurine).

(iii) **Gibberillins and Abscisic acid:** The GA3 is most commonly used in tissue culture. It promotes the growth of the cell culture at low density. Enhances callus growth and simulates the elongation of dwarf or stunted plantlets formed in culture. ABA in culture medium either stimulates or inhibits culture growth depending on species. It is most commonly used in plant tissue culture to promote distinct developmental pathways such as somatic embryogenesis.

5. Solidifying Agent

In the preparation of solid tissue culture medium, gel making and solidifying agents are commonly used. Agar (polysaccharide obtained from marine sea weeds) is used to solidify the medium. Normally 0.5-1 per cent agar is used in the medium to form a firm gel at the pH typical of plant tissue culture media. Use of high concentration of agar makes the medium hard and prevents the diffusion of nutrients into tissues.

pH

The adjustment of pH is essential for proper growth of plant cells and tissues The pH affects the uptake of ions, hence it must be adjusted between 5-6.0 by adding NaOH or HCl. Generally, the pH higher than six results in a fairly hard medium whereas pH below five does not allow satisfactory solidification of medium.

TABLE 11.2: Composition of Culture Medium Proposed by Murashige and Skoog [1962]

Sl.No.	*Nutrients*	*Formula*	*Quantity*
A	**Macro-nutrients**		
1	Ammonium Nitrate	NH_4NO_3	1.65 g
2	Potassium Nitrate	KNO_3	1.90 g
3	Calcium Chloride dihydrate	$CaCl_2.2\ H_2O$	0.44 g
4	Magnesium Sulphate hepta hydrate	$MgSO_4.7H_2O$	0.37 g
5	Potassium Dihydrogen Phosphate	KH_2PO_4	0.17 g
B	**Micro-nutrients**		
6	Ferrous Sulphate heptahydrite	$FeSO_4.7H_2O$	27.80 mg
7	Sodium Salt of EDTA	$Na_2EDTA\ 2H_2O$	33.60 mg
8	Potassium Iodide	KI	0.83 mg

Sl.No.	Nutrients	Formula	Quantity
9	Boric Acid	K_3BO_4	6.20 mg
10	Manganese Sulphate Heptahydrate	$MnSO_4$ $4H_2O$	22.30 mg
11	Zinc Sulphate Heptahydrate	$ZnSO_4$ $7H_2O$	8.60 mg
12	Sodium molybdate dihydrate	Na_2MoO_4 $2H_2O$	0.25 mg
13	Copper Sulphate Pentahydrate	$CuSO_4$ 5 H_2O	0.025 mg
14	Cobalt Chloride Hexahydrate	$CoCl_2$ $6H_2O$	0.025 mg
C	**Organic supplements**		
15	Myoinositol	Myoinositol	100.00 mg
16	Nicotinic acid	Nicotinic acid	0.05 mg
17	Pyridoxine HCl	Pyridoxine HCl	0.05 mg
18	Thiamine HCl	Thiamine HCl	0.05 mg
19	Glycine	Glycine	0.20 mg
20	Sucrose	Sucrose	20.00 mg
D	**Growth regulators as per need**		
	IBA		0.010 mg
	GA		0.100 mg
21	Gelling agent (only for solid medium)		
22	Agar	Agar	6-8 g/l
23	pH	pH	5.8

By making minor quantitative and qualitative changes a new media can be developed to suit specific requirements of explants.

PREPARATION OF NUTRIENT MEDIUM

There are two ways of preparing nutrient medium, *viz.*, from commercially available powder and from high quality chemicals. Now dry powdered media are commercially available. These media contains all the required nutrients. The powder is dissolved in distilled water than added to the medium and after adding sugar, agar and other desired supplements. The final volume is made up with distilled H_2O. The pH is adjusted and media is autoclaved.

Another method of preparing media is to prepare concentrated stock solutions by dissolving required quantities of chemicals of high purity in distilled water. Separate stock solutions are prepared for different media components such as major salts, minor salts, iron and organic nutrients except sucrose.

A separate stock solution is prepared for each growth regulator. All the stock solutions are stored in proper glass or plastic containers at low temperature in refrigerators. Stock solution of Iron is stored in amber colored bottles. Substances which are unstable in frozen state must be freshly added to the final mixture of stock solution at the time of medium preparation. Contaminated or precipitated stock solution should not be used.

STEPS INVOLVED

There are about ten sequential steps which are involved in the preparation of nutrient medium. These steps are listed as follows:

1. Appropriate quantity of Agar and sucrose is dissolved in distilled water.
2. Required quantity of stock solution, heat stable growth hormones (or) other substances are added by continuous stirring.
3. Additional quantity of distilled water is added to make final volume of the medium.
4. While stirring, the pH of the medium is adjusted by using 0.1 NaOH (or) HCl
5. If a gelling agent is used, heat the solution until it is clear.
6. The medium is dispensed into the culture tubes, flasks, (or) any other containers.
7. The culture vessels are either plugged with non-absorbent cotton wool rapped in cloth or closed with plastic caps.
8. Culture vessels are sterilized in autoclave at 121°C, 15Psi (1.06kg/cm^2) for about 15-20 min.
9. Heat labile constituents are added to the autoclaved medium after cooling to 30-40°C on a Laminar Airflow cabinet.
10. Culture medium is allowed to cool at room temperature and used or stored at 4°C (1 or 2 day).

SUMMARY

Explants are grown on a suitable artificially prepared nutrient medium called culture medium. A culture medium which contains important nutrients such as essential micro and macro nutrients, carbohydrate, vitamins and hormones is called nutrient medium. It is also known as culture medium. The pH of the nutrient medium should be 5.5.

There are two types of nutrient medium on which explants are grown. These are solid medium and liquid medium. Both media have some merits and demerits. Moreover, there are several differences between these two types of media. These have been briefly described.

The culture medium has to supply all the essential mineral ions required for *in-vitro* growth and morphogenesis of plant tissue. The culture medium consists of five main components, *viz.*, (i) inorganic nutrients [micro and macro], (ii) carbon source, (iii) organic supplements, (iv) growth regulators, and (v) solidifying agents. A brief description of all these components is presented. The composition of culture medium was first given by Murashige and Skoog (1962).

There are two ways of preparing nutrient medium, *viz.*, from commercially available powder and from high quality chemicals. The methods of preparing nutrient medium by both the ways have been explained. There are about ten sequential steps which are involved in the preparation of nutrient medium. These steps have been listed.

GLOSSARY

Liquid Medium: The tissue culture medium which is prepared by without using a solidifying or a gel making agent.

Macro-nutrients: Those elements which are required in concentration of 0.5 ml per litre.

Micro-nutrients: Those required in less than 0.5 ml per litre.

Nutrient Medium: A culture medium which contains important nutrients such as essential micro and macro are Nutrients, carbohydrate, vitamins and hormones.

Solid Medium: The tissue culture medium which is prepared by using a solidifying or a gel making agent.

QUESTIONS

1. **What are different types of culture medium? Describe their advantages and disadvantages.**
2. **Define solid nutrient medium. Describe its advantages and disadvantages.**
3. **Define liquid nutrient medium. Describe its advantages and disadvantages.**
4. **Explain briefly differences between solid and liquid nutrient medium.**
5. **Discuss briefly method of preparing nutrient medium.**
6. **Describe briefly various steps involved in the preparation of culture medium.**
7. **Define the following terms:**
   (i) Micro-nutrients (ii) Macro-nutrients
   (iii) Growth regulators (iv) Cytokinines
8. **Write short notes on the following:**
   (i) Solid culture medium (ii) Liquid culture medium
   (iii) Wet heat sterilization (iv) Dry heat sterilization
9. **Define nutrient medium and briefly describe composition of Murashige and Skoog (1962) culture medium.**

*****

Chapter 12

Tissue Culture Techniques

INTRODUCTION

Plant tissue culture refers to growth of living plant tissues in a suitable culture medium (*in vitro*). Culture medium is a nutrient medium which contains all essential micro and macro-nutrients, carbohydrates, vitamins and hormones. The pH of the nutrient medium should be 5.5. However, the culture medium differs from species to species. Thus a suitable medium has to be developed to meet the requirement of a plant species. In this technique, new plants are regenerated from tissues, cells and organs in test tubes in the nutrient medium. The organ includes any plant organ that has separate identity such as anther, ovule, embryo and bud. The plant part which is used for regeneration is called explant. It may be a cell, a protoplast, a tissue, or an organ. A mass of regenerated cells in culture medium is called callus (plural calli) and suspension of free cells of callus in a liquid medium is known as suspension culture. The regeneration capacity or ability of a plant cell to develop into a whole plant is known as totipotency, which reveals that each cell is capable of giving rise to a complete plant. The cell and tissue cultures lead to regeneration of complete plant. In some species like carrot, and sandalwood somatic embryos are developed but in several crops like wheat, rice, barley and tobacco development of both root and shoot takes place from the callus.

BASIC REQUIREMENTS

In tissue culture techniques, there are some basic requirements, *viz.* (1) aseptic conditions, (2) controlled temperature, (3) proper culture media, and (4) sub-culturing. These are briefly described below:

1. **Aseptic Conditions:** The tissue culture laboratory should have aseptic conditions. It means it should be well sterilized against pathogens. A pathogen free environment will help in maintaining good health of the callus, cell or protoplast cultures resulting in recovery of healthy plants

from such cultures. The explant and glassware should be properly sterilized before their entry into the tissue culture laboratory.

2. **Controlled temperature:** Air conditioning of the tissue culture laboratory is essential. Generally, temperature between 18-25°C is used. However, this varies from species to species. High temperature adversely affects the growth of the callus.
3. **Proper culture medium:** Culture media have been developed by various workers for different crop species. The medium has to be modified as per the requirement of a species. The culture media developed by Murashige and Skoog (1962) and Gamberg, *et al.*(1968) are used with some modifications in various crop species.
4. **Sub-culturing:** Transfer of tissue or callus from old culture media to fresh culture media is called sub-culturing. This is essential to maintain good health of the callus or tissues, because after some period, some nutrients are depleted in the culture media and change of media becomes essential.

IMPORTANT STEPS

Tissue culture technique generally consists of four important steps, *viz.* (1) isolation of tissues, (2) regeneration and callus formation in culture medium, (3) embryogenesis, and (4) organogenesis. These are briefly described below:

1. **Isolation of tissues:** Tissues for regeneration can be isolated with the help of sterilized blade from any plant part, *viz.* leaf, stem, apical bud, axillary bud, *etc.* The isolated tissues are sterilized and then grown on culture medium. Tissues should be isolated from disease free portion.
2. **Regeneration and callus formation:** Tissues proliferate on the culture medium and give rise to a mass of cells called callus. The callus is generally of two types, *viz.* friable callus and compact callus. The friable callus can be easily manipulated for suspension culture. However, compact callus is not suitable for suspension culture.
3. **Embryogenesis:** The process of formation of somatic embryos from the callus is called embryogenesis. Sometimes, somatic embryos are not formed rather somatic buds are formed which after germination give rise to a plant. Sub-culturing leads to healthy growth of callus and rapid embryogenesis.
4. **Organogenesis:** The process of differentiation of shoot and root from the somatic embryos is called organogenesis. Sometimes, a complete plant develops directly from the somatic bud. In such cases somatic embryos are not formed. Usually a plant develops from somatic embryos after germination. The plants thus obtained are transferred after sometime to pot culture from the culture medium. The soil of pots should be sterilized to make it pathogen free before transplantation of regenerated plants from culture medium to pots.

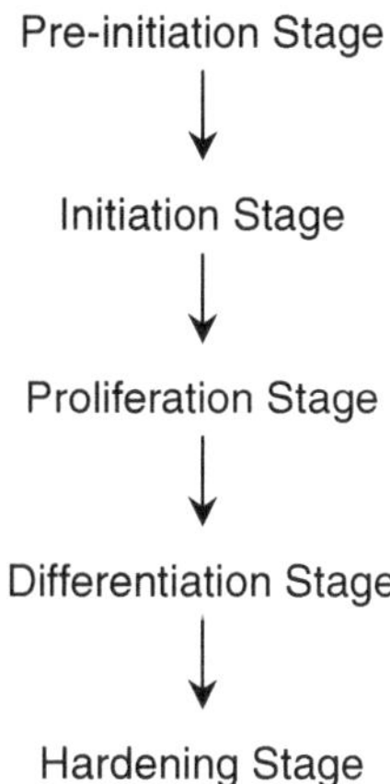

FIGURE 12.1: Stages of Plant Tissue Culture Technique.

TECHNIQUES OF TISSUE CULTURE

Tissue culture techniques are also referred to as *in vitro* techniques. *In vitro* techniques give rise to new plants by regeneration of tissues and organs in the nutrient medium. Various plant parts, *viz.* meristems, embryo, anther, cells and ovule are used for regeneration. Depending upon the plant part used as explant, plant tissue culture techniques are of five main types, *viz.*, (1) meristem culture, (2) embryo culture, (3) anther culture, (4) cell culture, and (5) ovule culture. A brief description and practical applications of these techniques in crop improvement are presented below:

1. Meristem Culture

Regeneration of whole plant from tissues of an actively dividing plant part such as stem tip, root tip or auxillary bud is called meristems culture. Generally shoot apical meristem is used for regeneration. In other words, tissues of shoot apex are used for culturing in the nutrient medium. Regeneration is obtained in a suitable culture medium. The medium differs from species to species. This technique has been widely used in vegetatively propagated crops such as sugarcane, potato, banana and several fruit trees and timber species. The main applications of meristem culture in crop improvement are given below:

1. It is used for micro-propagation (Mass production of clonal progeny through tissue culture) in banana, strawberries, citrus and some timber trees such as *Delbergia sissoo* (sheesham).
2. Virus free plants can be obtained through this technique, because meristematic cells are almost free from virus even in the virus infested plants.

3. The exchange of germplasm of plantlets obtained by meristem culture is safe, because such material is free from insect and pathogens. This is useful in the exchange of germplasm of asexually propagated plant species.
4. The germplasm can be conserved in the form of meristems at –196°C for long term storage in the liquid nitrogen. In other words, meristems are suitable for cryopreservation.

2. Embryo Culture

Regeneration of whole plant from an embryo in the culture medium is called embryo culture. Embryos of appropriate stage are removed from the seed and are transferred to culture medium. This technique is used when there is dishamony between the embryo and endosperm. This technique is used to make distant crosses successful. The embryo in distant crosses is removed before abortion and cultured in nutrient medium. This technique has been used to make interspecific hybridization successful in *Trifolium, Lycopersicon* and several other geners. It has also been used to make intergeneric crosses successful between *Hordeum* and *Secale, Hordeum* and *Triticum, Triticum* and *Secale, Triticum* and *Aegilops, etc.* The main uses of embryo culture technique in crop improvement are given below:

1. It is useful in making distant crosses (Interspecific and intergeneric) successful.
2. It is useful in obtaining haploid plants from interspecific and intergeneric crosses. Hoploids have been obtained from intespecific crosses in barley and also from intergeneric crosses between wheat and barley.
3. In orchids, seeds lack stored food and are unable to propagate. Embryo culture is useful in getting seedlings in such species. Thus embryo culture is useful in orchid breeding.
4. In some plant species seeds have long dormancy period *i.e.* upto two years. Such dormancy is found in *Prunus, Iris* and *Taxus*. In these species, we can get seedling immediately by embryo culture. Thus embryo culture is effective in breaking long seed dormancy.

3. Anther-Culture

Regeneration of whole plant from another or pollen in the culture medium is called anther or pollen culture. Anthers of appropriate stage are taken from the plant. The optimum stage may differ from species to species. This technique is used to obtain haploid plants. This technique has been used in different crops such as *Brassica,* tobacco, *Petunia,* rice, barley, wheat, tomato, potato, *Triticale, etc.* Main uses of this technique in crop improvement are given below:

1. It is useful in development of haploids. Haploids have been obtained by pollen culture in wheat, barley, rice, *Triticale,* potato, *Brassica campestris, etc.*
2. By doubling the chromosome number of haploids we get homozygous diploids.

4. Cell Culture

Regeneration of whole plant from callus and suspension cultures in the nutrient medium is known as cell culture. Protoplasts (naked cells or cells without cell wall) are also used for regeneration of whole plants. Cell culture technique has been used in different crop plants such as wheat, rice barley, maize, tobacco and some fruits and timber trees. The main uses of this technique are given below:

1. It is useful in mass clonal production of different species.
2. Protoplast culture is useful in somatic hybridization and overcoming the barriers of cross incompatibility in interspecific and inter-generic crosses.
3. Cell culture and protoplast cultures are useful in the development of transgenic (genetically engineered) plants. This technique is being used for genetic transformation in almost all important crop plants.
4. Embyogenic cell cultures are suitable for cryopreservation, *i.e.*, preservation in liquid nitrogen at –196°C. Embryogenic cultures are used for long term preservation in many crop species.

5. Ovule-Culture

Regeneration of whole plant from the ovule in the nutrient medium is called ovule culture. This is also known as ovary culture. Two types of ovules, *viz.*, (1) unfertilized, and (2) fertilized may be available. This technique is used for embryo rescue in distant crosses. Haploid plants are obtained from culture of unfertilized ovules and diploid from the fertilized ovules. This technique has been used to limited extent.

APPLICATIONS IN CROP IMPROVEMENT

Tissue culture has several useful applications in crop improvement. The main applications are: (1) generation of variability (2) development of haploids, (3) embryo rescue, (4) somatic hybridization, (5) selection for disease resistance, (6) selection for salinity and metal toxicity resistance, (7) selection for drought resistance, (8) micro-propagation, and (9) preservation of germplasm. These aspects are briefly described as follows:

1. Induction of Variability

The variation which is induced in the tissue culture is an important source of variability for crop improvement. Depending upon the explant, tissue culture induced variation is of three types, *viz.*, gametoclonal, somaclonal and protoclonal.

(*i*) **Gametoclonal variation:** The variation which is observed among the plants which are regenerated from gametic culture is called gametoclonal variation. Such variation is observed among the plants which are regenerated from anther or ovule culture.

(*ii*) **Somaclonal variation:** The variation that is observed among the plants which are regenerated from callus cultures of somatic explants such as meristems is called somaclonal variation.

(*iii*) **Protoclonal variation:** The variation which is observed among the plants which are regenerated from callus cultures of protoplast is referred to as protoclonal variation. Main features of culture induced variation are briefly presented below:

(1) The variation is genetic in origin and hence heritable.

(2) Somaclonal variation occurs in both sexually and asexually propagated species. But the frequency of such variation is very high in vegetatively propagated species (up to 75 per cent in potato and sugarcane) and low in seed propagated species (0.8-1.2 per cent in maize).

(3) It occurs in both oligogenic as well as polygenic characters.

(4) In majority of cases, somaclonal variation is heterozygous in origin. Rarely the variation is true breeding or homozygous as has been reported in wheat and mustard somaclones.

(5) All three types of culture variations result due to chromosomal changes such as deletion, duplication, inversions and translocations.

(6) The frequency of variation is higher in protoclones than in somaclones and gametoclones due to stress imposed by the process of cell wall removal and its subsequent synthesis.

Somaclonal variation is useful in several ways. It helps in the (1) isolation of disease resistant, (2) salt tolerant, (3) herbicide tolerant, (4) metal toxicity tolerant, and (5) early maturing variants. Disease resistant somaclones have been isolated in several crops like sugarcane, potato, tobacco and ornamental plants. In many crops, heritable somaclonal variations lead to the development of new germplasm. From such variation mutants with resistance to diseases, stress conditions and good quality can be isolated after proper screening.

2. Development of Haploids

Haploids can be developed by tissue culture technique. Haploids are developed by anther culture. Haploids are made diploid by colchicine treatment. Haploids have been developed in more than 250 plant species by anther-culture technique. In China, one variety each in wheat [Jinghua 1] and rice [Guan 18] have been developed for commercial cultivation by tissue culture. The new varieties are better in agronomic properties than old varieties.

3. Embryo Rescue

Embryo culture technique helps in making the interspecific crosses successful when there is post fertilization disharmony between the embryo and endosperm. The embryo in interspecific cross is removed before abortion and cultured in nutrient medium. Embryo rescue technique has been used for successful interspecific hybridization in genus *Trifolium* and *Lycopersicon*. Embryo culture technique has also been used for making intergeneric hybrids successful between *Triticum* and *Aegilops, Triticum* and *Secale* and several other genera.

In barley, isolation of ovules from interspecific crosses immediately after fertilization has helped in regeneration of barley plants in Germany. At International

Rice Research Institute, Philippines, embryo culture technique has been successfully used for transfer of Brown Plant Hopper (BPH) resistance from wild species *Oryza officinalis* to the cultivated species *O. sativa*. Similarly, embryo rescue technique has been used in several other crops. The main drawback of embryo culture is that it is applicable to those distant crosses where crossing is possible and problems are related to post fertilization.

4. Somatic Hybridization

Crossing of plants through fusion of somatic cells is known as somatic hybridization. In such hybridization, the sexual process is bypassed. The fusion of cells takes place through protoplasts. Protoplasts are naked cells or cells without cell wall. Somatic hybridization through protoplast fusion permits hybridization between any two plants irrespective of their taxonomic relationship. In other words, it makes incompatible crosses possible. Somatic hybridization consists of several steps, *viz.* (1) isolation of protoplasts, (2) fusion of protoplasts, (3) selection of hybrid cells, (4) culture of hybrid cells, (5) regeneration of plants from hybrid tissues, and (6) characterization of hybrid plants.

5. Selection for Disease Resistance

For selection of disease resistant lines, the pathogen is included in the culture medium. The callus cultures of susceptible genotypes support the growth of pathogenic fungi while resistant cells do not support as has been observed in potato and tobacco. In potato, resistant plants to wilt and root rot caused by *Fusarium oxysporium* have been isolated by tissue culture technique. In tomato, tobacco and alfalfa, resistance to virus was obtained by infection of callus by weak virus. Tissue culture can help in reviving some of the varieties of crop plants which have become obsolete due to their susceptibility to a particular disease. Mosaic virus is a serious problem in pulse crops like greengram, blackgram, soybean and other crops like papaya and okra. Tissue culture can help in isolation of mosaic free lines in these crops. Disease free material has been developed in potato, tobacco, tomato, sugarcane and some fruit crops through tissue culture technique.

6. Selection for Salinity Resistance

Tissue culture plays an important role in the identification and isolation of salt resistant genotypes. The cell culture technique is used in breeding of salinity resistant genotypes. Millions of cells are subjected to an elevated level of saline solution in a flask. Only resistant or tolerant cells will survive in such solution. The surviving cells are used for regeneration of salinity resistant plants. Thus, cell culture is a simple and rapid method of developing salinity resistant genotypes. Similarly, genotypes resistant to herbicides and metal toxicity can be isolated.

7. Selection for Drought Resistance

In the culture medium drought is induced by the use of a chemical known as PEG (Polyethylene glycol). Drought resistance or water stress tolerant genotypes have been isolated through tissue culture technique in tomato and *Sorghum*. Low proline content is an indicator of water stress tolerance. It also helps in isolation of water stress tolerant lines in tissue culture.

8. Micro-propagation

Tissue culture technique is useful in rapid and mass multiplication of plant material. Micro-propagation is used for this purpose. Micro-propagation refers to regeneration of plants from isolated meristemetic cells or tissues or from somatic cells. It is also known as micro-cloning. Micro-propagation can be used for mass multiplication of crop plants which are difficult to propagate sexually or those vegetatively propagated species in which rate of multiplication is slow. Micro-propagation can also be used for mass multiplication of superior hybrids as an alternative to the production of hybrid seed. Micro-propagation has several advantages as given below:

1. It ensures pathogen free status of the propagules. Propagules are small plants developed by micro-propagation.
2. It helps in rapid multiplication of material. Large number of propagules can be obtained from a single plant by this method.
3. The material multiplied by this technique can be maintained in a small place. The transportation of such propagules from one place to another is also easy.
4. The micro-propagation can yield results faster than conventional breeding. It is mostly used in horticulture, floriculture and forestry.

This technique has been used to obtain early and disease free lines in strawberries, banana, citrus and some timber trees. In field crops, micro-propagation has been used in potato and sugarcane. The main problems with this technique are : (1) somaclonal variation, and (2) variation in chromosome number due to continuous sub-culturing.

9. Germplasm Preservation

Preservation of germplasm in the form of tissues is an important application of tissue culture in crop improvement. The cell or tissues can be preserved in liquid nitrogen in the long term storage. The cells or tissues are treated with dimethyl sulphoxide to protect against freezing injury. Preservation of tissues is more useful in vegetatively propagated crops which usually are unable to produce seed. Such storage would require less space.

SUMMARY

Plant tissue culture refers to growth of living plant tissues in a suitable culture medium (*in vitro*). Culture medium is a nutrient medium which contains all essential micro and macro nutrients, carbohydrates, vitamins and hormones.

In tissue culture techniques, there are some basic requirements, *viz.* (1) aseptic conditions, (2) controlled temperature, (3) proper culture media, and (4) sub-culturing. Tissue culture technique generally consists of four important steps, *viz.*

(1) isolation of tissues, (2) regeneration and callus formation in culture medium, (3) embryogenesis, and (4) organogenesis.

Depending upon the plant part used as explant, plant tissue culture techniques are of five main types, *viz.*, (1) meristem culture (2) embryo culture, (3) anther culture, (4) cell culture, and (5) ovule culture.

Tissue culture has several useful applications in crop improvement. The main applications are: (1) generation of variability (2) development of haploids, (3) embryo rescue, (4) somatic hybridization, (5) selection for disease resistance, (6) selection for salinity and metal toxicity resistance, (7) selection for drought resistance, (8) micro-propagation, and (9) preservation of germplasm.

GLOSSARY

Anther or pollen Culture: Regeneration of whole plants from anther or pollen in culture medium.

Embryo Culture: Regeneration of whole plant from an embryo in the culture medium.

Embryo-genesls: The process of formation of somatic embryos from the callus.

Meristem or Callus Culture: Regeneration of whole plant from tissues of actively dividing plant parts such as stem tip or apical bud, root tip or auxillary.

Micro-propagation: Regeneration of plants from isolated meristemetic cells or tissues.

Organogenesis: The process of differentiation of root and shoot from somatic embryos.

Ovule or Ovary Culture: Regeneration of whole plant from the ovule in nutrient medium.

Plant tissue culture: Growing of living plant cells, tissues or organ on a suitable nutrient medium outside the tissues of parent plant.

Somaclonal Variation: The variability generated by the use of tissue culture.

Suspension Culture: Regeneration of whole plant from single cell in liquid culture medium.

QUESTIONS

1. **List different techniques of plant tissue culture. Describe role of any two techniques in crop improvement.**
2. **What is tissue culture? Describe its role in crop improvement.**
3. **Define anther culture and embryo culture. Discuss their application in crop improvement.**
4. **What is meristem culture? Explain briefly its role in plant breeding.**
5. **Explain briefly applications of cell culture in crop improvement.**

6. **Describe briefly basic requirements and various steps involved in tissue culture technique.**

7. **Write short notes on the following:**
   (*a*) Micro-propagation (*b*) Somatic hybridization
   (*c*) Somaclonal variation (*d*) Tissue culture
   (*e*) Sub-culturing (*f*) Culture medium

8. **What do the following terms signify?**
   (*a*) Explant (*b*) Totipotency
   (*c*) Organogenesis (*d*) Transgenic plants
   (*e*) Callus (*f*) Cryopreservation

9. **Differentiate between the following:**
   (*a*) Organogenesis and embryogenesis
   (*b*) Anther culture and meristem culture
   (*c*) Cell culture and protoplast culture
   (*d*) Somaclones and gametoclones.

10. **Explain briefly the following:**
    (*a*) Embryo rescue (*b*) Anther culture
    (*c*) Meristem culture (*d*) Cell culture
    (*e*) Protoplast culture (*f*) Ovule

*****

Chapter 13

Micro-propagation

INTRODUCTION

Mass production of clonal progeny through tissue culture technique is referred to as micro-propagation. In other words, *in vitro* mass production of clonal progeny is called micro-propagation. The main points related to micro-propagation are listed as follows:

In micro-propagation, tissues from actively dividing plant parts such as stem tip, root tip or auxillary buds are used. The use of shoot apical meristem is very common for micro-propagation. The explants are selected from disease free portion of parent plants.

The mass production of clonal plants is carried in culture medium or nutrient medium. The nutrient medium slightly differs from species to species.

STEPS INVOLVED

The process of Micro-propagation consists of four distinct stages, *viz.*, (i) selection of suitable explants, (ii) proliferation of shoots from the explants on medium, (iii) transfer of shoots to a rooting medium, and (iv) transfer of plants to soil or normal environment. These are briefly discussed as follows:

1. Selection of Suitable Explants

Selection of suitable and well sterilized explants is essential for establishment of contamination free cultures. Cultures are initiated from various type of explants such as meristem shoot tips, nodal buds, internodal segments leaves, young inflorescence are used for initiation of meristem culture. However, meristem, shoot tips and nodal buds are preferred most for commercial micro-propagation. Meristems < 0.2mm (or) 0.2-0.4mm are devoid of pathogens and thus result in the production of virus free plants through micro-propagation Selected explants are surface sterilized and aseptically cultured on a suitable medium. During this stage a simple medium without hormones and with low levels of both auxins and cytokinins is used. In general the explants taken from Juvenile plants respond better. Cultures are incubated in a room maintained at 25 + 1°C temperature 16/8hr light/dark 3000-5000 lux light intensity and 50-70 per cent relative humidity. After 2-4 weeks of incubation, the effective cultures resume their growth.

2. Proliferation on Medium

This stage consists of multiplication of propagules. Effective explants from stage I are sub-cultured on a fresh culture medium. The time of application and concentration of auxins and cytokinins in multiplication medium are important factors that affect the extent of multiplication. *In vitro* multiplication of shoots can be achieved in four ways, *viz.*, by callus culture, adventitious shoots, apical and axillary shoots and somatic embryo genesis. These are briefly explained as follows:

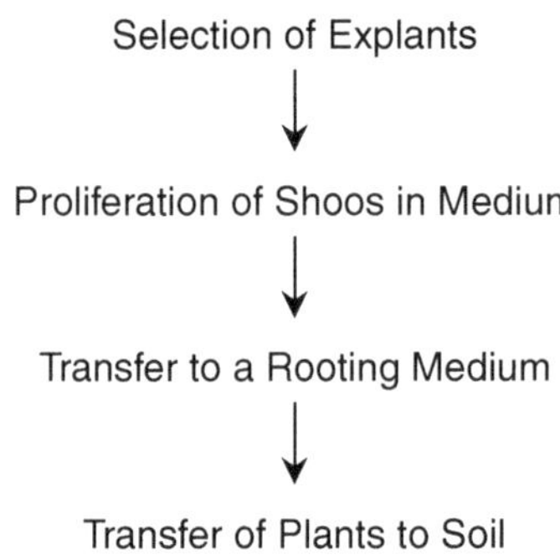

FIGURE 13.1: Steps involved in Micro-propagation.

(i) Multiplication through Callus Culture

A large number of plantlets can be obtained from callus either through shoot and root formation or somatic embryogenesis. Mass production of callus followed by shoot regeneration would be ideal method for large-scale propagation of desired plants but there are two main drawbacks of this method. Firstly, repeated sub-culturing leads to reduction I shoot formation. Secondly, it leads to production of aneuploids and polyploids. Due to these drawbacks, multiplication of shoots through callus culture is less preferred. However, this approach has been successively used in citrus and oil palms.

(ii) Multiplication by Adventitious Shoots

Adventitious shoots are stem and leaf structures which develop naturally on plant at places other than normal leaf axis regions. These structures include stems, buds, tubers, corms, rhizomes *etc.* In many horticultural crops, vegetative propagation through adventitious bud formation is in commercial practice in *Begonia* and some other ornamental plants. In culture also similar type of adventitious shoot formation can be induced by using appropriate combination of growth regulators in media. Adventitious buds can also be induced on the leaf and stem cuttings of even those species which are normally not propagated vegetative such as flax and *Brassica* species. Development of adventitious shoots directly from excised organs is preferred for clonal propagation as compared to callus formation method because in the former diploid individuals are formed but in the later often cytologically abnormal plants are produced. However in varieties which have genetic chimeras, multiplication through adventitious bud formation may lead to splitting the chimeras to pure type of plants.

(iii) Multiplication by Apical and Axillary Shoots

Apical shoots are those that occupy growing tip of shoots whereas axillary shoots are those that develop from their normal positions on the plant in a axis of each leaf. Every bud has got potentiality to develop into a shoot. Apical dominance plays an important role in the development of axillary buds and is governed by the growth regulators. In the species where the apical dominance is very strong removal or injury of terminal bud is necessary for the growth of axillary bud. To initiate culture, shoot tips are placed on the medium containing low levels of cytokinins (0.05 to 0.5 mg/lit BAP) and auxin (0.01 to 0.1mg/lit, IBA). The level of cytokinins is progressively raised at each subculture, until the desirable rate of proliferation is achieved. In cultures, the rate of shoot multiplication can be enhanced by culturing the shot tips on the medium containing cytokinins. In such cases, the cultured explants transform into the mass of branches. By further sub-culturing, shoot multiplication cycle may be repeated and culture may be maintained round the year and proliferating shoots of many species have been maintained up to 10-15 years. The enhanced axillary branching method of shoot multiplication may be initially slower than the other two methods. This method is popular for clonal propagation of crop plants because the cells of shoot apex are uniformly diploid and are least susceptible to genetic changes.

(iv) Multiplication by Somatic Embryogenesis

Multiplication by somatic embryogenesis in nature is generally restricted to intra ovular tissues. For example, any cell of gametophytic or sporophytic tissue around the embryo sac, cells of nucellus or integument of members of the family Rutaceae can develop into embryos. However, in plant tissue culture embryos can develop even from somatic cells like epidermis, parenchymatous cells of petioles or secondary phloem, *etc.* Somatic embryos are formed either by direct embryogenesis or by indirect embryogenesis (Sharp *et al.*1980). In the former case, embryos are formed directly from explant tissue in the absence of callus proliferation. In the latter case, first callus is formed from explants from which embryos are formed. Somatic embryogenesis as a means of micro-propagation is seldom used because this method is little difficult, leads to occurrence of mutations, associated with seed dormancy and difficulty in embryogenesis.

3. Transfer of Shoot to a Rooting Medium

Plantlets with shoots proliferated during stage II are transferred to a rooting medium. In general rooting medium has low salt. All cytokinins inhibit rooting. In most species 0.1 mg/lit NAA and IBA is required for rooting. The availability of IBA induces primary/secondary roots where as NAA induces root hairs. Generally, plantlets with shoots size of approximately 2 cm are transferred to rooting medium. Plantlets with 0.5 –1cm roots are usually transplanted into pots since longer roots tend to get damaged during the transfer.

4. Transfer of Plants to Soil/Normal Environment

This is the most important step in micro-propagation. The ultimate success of micro-propagation on commercial scale depends on their survival rates when transferred to the soil. Inside the greenhouse the plants increase their resistance to moisture stress and disease. The plantlets have to become autotrophic in contrast to their heterotrophic state induced in micro-propagation culture. Transfer of plantlets to soil is the most crucial step in micro-propagation. The plantlets are maintained under highly protected conditions in *in vitro, i.e.,* high humidity, low irradiance, low CO_2 levels and high sugar content.

SHOOT MERISTEM CULTURE

Meristem culture refers to cultivation of axillary or apical shoot meristems, particularly of shoot apical meristem in culture medium. The shoot apical meristem is the portion lying distal to the youngest leaf primordium, it is upto about 100 μm in diameter and 250 μm in length. The first application of meristem culture was to obtain virus-free plants of dahlias. In 1952, Morel and Martin isolated 100 μm long meristem from virus -infected plants, and cultured them to obtain virus free shoots. Since then the technique of meristem culture has been greatly refined and used for obtaining plants free from viruses, viroids, mycoplasma and even fungi and bacteria in a range of crops.

In India, some valuable clones of potato, sugarcane, *etc.* free from virus have been obtained from virus infested plants through meristem culture. Care must be taken to remove the apical meristem with as little surrounding tissue as possible to minimize the chances of virus particles being present in the explants. This application of meristem culture is of great value, particularly in the maintenance of breeding materials and germplasm exchange, which are invaluable for any breeding program.

PROCEDURE OF MERISTEM CULTURE

1. **Isolation of Apical Meristem:** Considerable expertise is required to dissect out the shoot apical meristem with only one or two leaf primordia (100-500 μm in length). Care has to be taken to prevent desiccation, and contamination by the virus present in the surrounding tissue.
2. **Growth Regulators:** Generally, growth regulators (usually, small amounts of an auxin and a cytokinin) are added to the medium to support shoot growth from the cultured meristems.
3. **Size of Explants:** In general, the larger the meristem explant, the greater the chances of its survival and shoot development. But the risk of infection by the virus also increases with larger explants. Therefore, a compromise

has to be reached between these two opposing forces in deciding the size of explants.

4. **Thermotherapy:** Viruses are eliminated by ***thermotherapy*** of whole plants, in which plants are exposed to temperatures between 35-40°C for a few minutes to several weeks depending on the host-virus combination. Thermotherapy is often combined profitably with meristem culture to obtain virus free plants. In general, shoot-tips are excised from heat-treated plants, but cultured meristems may also be given thermotherapy; the former is preferable since larger explants can be safely taken from heat-treated plants.
5. **Low Temperature Treatment:** A prolonged exposure to a low temperature (5°C), followed by shoot-tip culture, has proved quite successful in virus elimination; this is often called ***cryotherapy.***
6. **Chemical Treatment:** Some chemicals, *e.g.*, virazole (ribavirin), cucloheximide, actinomycin D, *etc.*, which interfere with virus multiplication, may be added into the culture medium for making the shoot-tips free from Viruses; this is known as ***chemotherapy.***

ADVANTAGES AND APPLICATIONS

There are several advantages of micro-propagation such as high multiplication rate, rapid method, maintenance of desirable lines, alternative to mass production, useful with dormancy, virus free plants, selective propagation, year round propagation, exchange of material, germplasm conservation, *etc.* These are discussed as follows:

1. **High Multiplication Rate:** Micro-propagation of a hybrid has the greatest multiplication advantage since it can result in large number of elite plants from a very small portion of tissues taken from the hybrid plant.
2. **Uniform Plants:** It is useful to get genetically uniform plants in large number. Few explants are enough to produce millions of plants with extremely high multiplication rate.
3. **Rapid Method**: Rapid multiplication of rare and elite genotypes. There is considerable reduction in period between selection and release of new variety.
4. **Maintenance of Desirable Lines**: Micro-propagation is useful in maintenance of desirable lines such as inbred lines and male sterile lines for producing commercial F_1 hybrids. It is also useful in multiplication of particular heterozygous superior genotype with increased productivity such as in oil palm.
5. **Alternative to Mass Production**: This technique is possible alternative in plants species which do not respond to conventional bulk propagation technique.
6. **Useful with Dormancy**: In plants with long seed dormancy micro-propagation is faster than seed propagation.

7. **Virus Free Plants:** It is useful to obtain virus free stocks. Virus elimination generally improves the yield by 20-90 per cent over infected controls. Virus-free plants serve as excellent experimental materials for evaluating the detrimental effects of infections by various viruses. Meristem culture can also help eliminate other pathogens, *e.g.*, mycoplasmas, bacteria and fungi. Bacteria and fungi present in explants show up when they are cultured *in vitro* since tissue culture media provide excellent nutrition for the microbes.

   Meristem culture has been used to eliminate systemic bacteria from *Diffenbachia* and *Pelargonium*, and *Fusareum roseus* from carnations.
8. **Selective Propagation**: In dioecious species plants of one sex is more desirable than those of other sex such as male asparagus and female papaya. In such cases plants of desired sex can be selectively multiplied by this technique.
9. **Year Round Production**: This technique is carried out throughout the year independent of seasons.
10. **Exchange of Material**: Facilitates speedy international exchange of plant materials.
11. **Germplasm Conservation:** Meristems have been identified as an excellent material for germplasm preservation of some species.
12. **Better Quality:** In case of ornamentals tissue culture derived plants give better growth, more flowers and less fall out.
13. **Requires less Space:** Tissue culture can be used to minimise the growing space in commercial nurseries for maintenance of shoot plant.

LIMITATIONS

There are some limitations of micro-propagation such as high cost, standardization of technique, somaclonal variation, browning of medium, high technical skill, difficulty in hardening, occurrence of off types, possible at small scale, *etc.* These are discussed as follows:

1. **High Cost:** This technique has limited application because of high production cost.
2. **Standardization of Technique:** At each stage the technique has to be standardized. Suitable techniques of micro-propagation are not available for many crop species.
3. **Somaclonal Variation:** Somaclonal variation may arise during *in vitro* culture especially when a callus phase is involved, *e.g.*, banana
4. **Browning of Medium**: Browning of medium is a problem in some trees and perennial plants. The browning is caused by secretion of certain chemicals. Higher concentration of such compounds may cause death of the explants.

5. **High Technical Skill**: This technique requires highly technical skills.
6. **Susceptibility of Plants**: The plantlets are susceptible to water losses in external environment and they have to be hardened to the external atmosphere.
7. **Difficulty in Hardening**: Plants requires a transitional period before they are capable of independent growth. The plants obtained have low photosynthetic rate and are not self sufficient. Acclimatization is difficult process to get high percentage of survival of plants.
8. **Occurrence of off types**: Continuous propagation from same material for many generations may lead to occurrence of many off types in the culture.
9. **Possible at Small Scale**: This technique is suitable for small-scale production of clonal progeny. It is less suitable for commercial scale.
10. **Chances of Contamination:** Despite precautions during culturing there are chances of contamination by various pathogens which may cause vary high losses in a short time.

SUMMARY

Mass production of clonal progeny through tissue culture technique is referred to as micro-propagation. In other words, *in vitro* mass production of clonal progeny is called micro-propagation.

The process of micro-propagation consists of four distinct stages, *viz.*, (i) selection of suitable explant, (ii) proliferation of shoots from the explants on medium, (iii) transfer of shoots to a rooting medium, and (iv) transfer of plants to soil or normal environment. These have been discussed.

Meristem culture refers to cultivation of axillary or apical shoot meristems, particularly of shoot apical meristem in culture medium. The shoot apical meristem is the portion lying distal to the youngest leaf primordium, it is up to about 100 µm in diameter and 250 µm in length.

There are some problems which are associated with micro-propagation such as microbial contamination, callus formation, somaclonal variation, browning of medium, vitrification and transplantation shock. These have been discussed.

There are several advantages of micro-propagation such as high multiplication rate, rapid method, maintenance of desirable lines, alternative to mass production, useful with dormancy, virus free plants, selective propagation, year round propagation, exchange of material, germplasm conservation, *etc.* These have been discussed.

There are some limitations of micro-propagation such as high cost, standardization of technique, somaclonal variation, browning of medium, high technical skill, difficulty in hardening, occurrence of off types, possible at small scale, *etc.* These have been discussed.

GLOSSARY

Germplasm: Sum total of genes in a breeding population.

Meristems: Actively dividing plant cells or tissues.

Micro-propagation: Mass production of clonal progeny through tissue culture technique. In other words, mass production of miniature planting material through *in vitro* techniques.

Propagules: Miniature planting material.

Vitrification: The brittle, glassy and water soaked appearance of some shoots in culture medium.

QUESTIONS

1. **Define micro-propagation and discuss important points related to this technique.**
2. **Describe briefly various steps involved in micro-propagation.**
3. **Define shoot meristems. Describe briefly procedure of shoot meristem culture.**
4. **Discuss briefly various problems associated with micro-propagation.**
5. **Give a brief account of advantages and applications of micro-propagation.**
6. **Explain various limitations of micro-propagation.**
7. **Write short notes on the following:**
   (*a*) Meristem culture
   (*b*) Advantages of micro-propagation

*****

Chapter 14

Synthetic Seeds

INTRODUCTION

Artificial seeds, which are also known by other names such as "synseeds", were first described by Murashige in 1977. He defined artificial seeds as an encapsulated single somatic embryo. An artificial seed was later defined by Gray *et al.*1991, as a somatic embryo that is engineered for the practical use in commercial plant production. Thus, artificial seeds or synthetic seeds are the living seed like structure containing somatic embryos (derived from plant tissue culture) encapsulated by a hydrogel. They behave like true seeds if grown in soil and can be used as a substitute of natural seeds. It can be called as artificial seed.

The first synthetic seeds were produced by Kitto and Janick in 1982 using carrot. In simple words, synthetic seed contains an embryo produced by somatic embryogenesis enclosed within an artificial medium that supplies nutrients and is encased in an artificial seed covering. The concept of artificial seeds was then limited to those plant species in which the production of their somatic embryos is possible.

STEPS INVOLVED

The important steps involved in the production of synthetic seeds include selection of explants, induction of callus, induction of somatic embryos, proliferation of embryos, maturation of embryos, desiccation and tolerance induction, encapsulation and *in vitro* germination or transplantation in the field for germination. These steps are briefly explained as follows:

1. **Selection of Explant:** The explant is selected from the material of interest. Usually apical meristem tissues are used for this purpose. The explants should be carefully sterilized before entry in the tissue culture laboratory.
2. **Induction of Callus:** The selected explants are transferred to nutrient medium for induction of callus. The culture medium slightly differs from species to species for callus induction. Generally, Murashige and Skoog culture medium is used for callus induction.

3. **Induction of Somatic Embryos:** The somatic embryos are inducted from callus by *in vitro* techniques. Growth regulators may added in the culture medium, if need, for somatic embryogenesis.

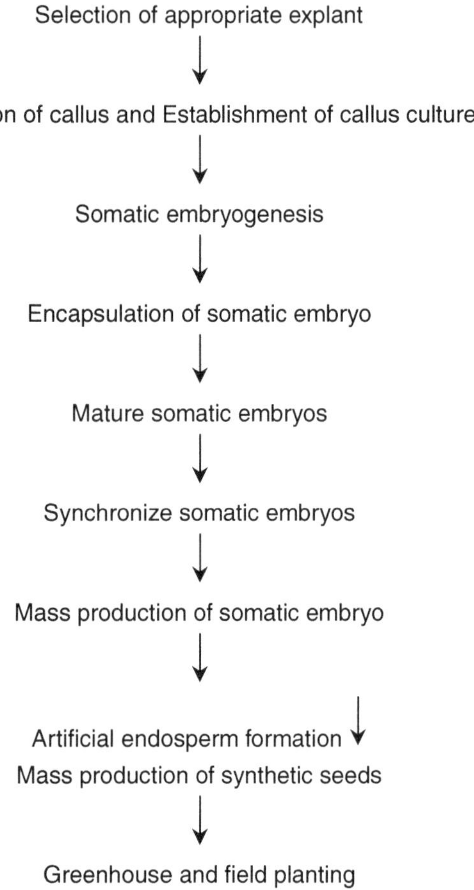

FIGURE 14.1: Steps in Artificial Seed Production.

4. **Proliferation of Embryos:** The proliferation of somatic embryos is achieved using tissue culture techniques. Sub-culturing is done periodically for proliferation.
5. **Maturation of Embryos:** The next step is histo-differentiation and maturation of somatic embryos. Maturation of somatic embryos is essential to get higher rate of germination. Somatic embryo may be hardened either by treating or coating mature somatic embryos with a suitable polymer followed by drying or treating them with abscisic acid [ABA] during their maturation phase. The ABA treatment also improves germination of somatic embryos.

6. **Tolerance Induction:** The tissue culture induced somatic embryos are delicate to external conditions. Hence, their desiccation and hardening is essential before encapsulation with
7. **Encapsulation:** Synthetic seeds are encapsulated with the nutrients, growth regulators, pesticides, antibiotics, *etc.* The desiccated synthetic seeds are encapsulated with a suitable polymer, whereas hydrated synthetic seeds are encapsulated with a suitable gel. Encapsulation of synthetic seeds is done to provide them protection against insects and pathogens.

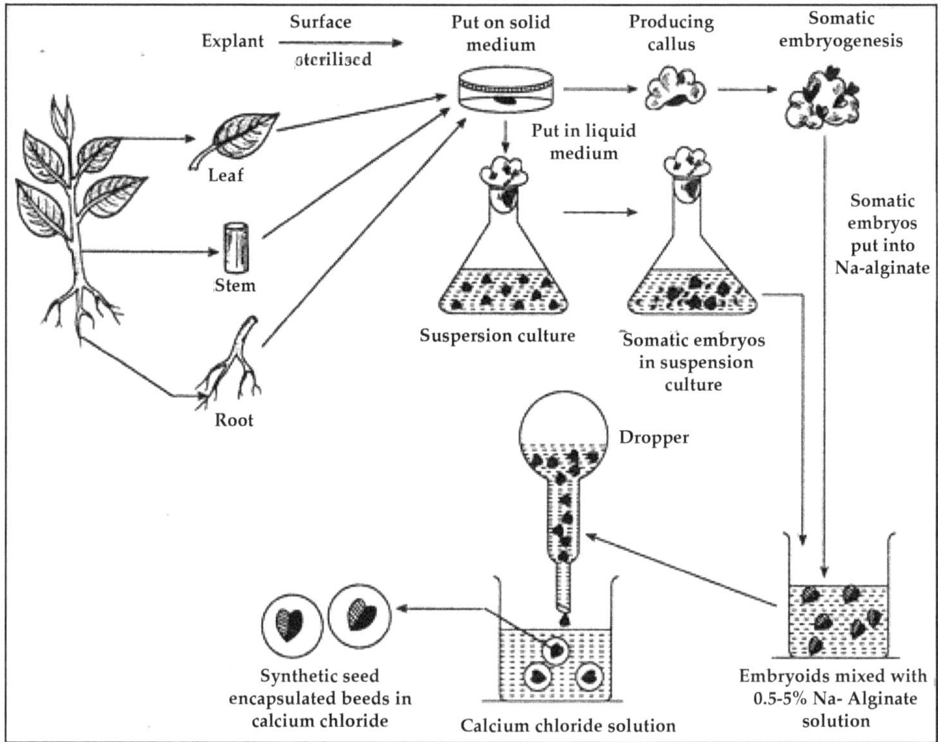

FIGURE 14.2: Steps for Synthetic or Artificial Seed Production.

8. **Germination:** This is the final step in synthetic seed production. The germination test is conducted both under *in vitro* conditions as well as under field conditions to assess the viability of somatic embryos.

METHOD OF PRODUCTION

Synthetic seeds are generally produced in two ways, *viz.*, (i) desiccated, and (ii) hydrated systems. These are discussed as follows:

1. **Desiccated System:** In the desiccated system the somatic embryos are first hardened to withstand desiccation and then are encapsulated in a suitable coating material to yield desiccated artificial seeds. Somatic embryo may be hardened either by treating or coating mature somatic embryos with a

suitable polymer followed by drying or treating them with abscisic acid [ABA] during their maturation phase. The ABA treatment also improves germination of somatic embryos.

2. **Hydrated System:** In this system, various water soluble gels can be used for the purpose of encapsulation, like sodium alginate, potassium alginate, carrageenan, sodium pectate or sodium alginate with gelatin. Among the gels, calcium alginate is the most suitable and commonly used. There are some problems of hydrated seeds as listed below:
   (*a*) Hydrated seeds are sticky and difficult to handle on a large scale.
   (*b*) They dry rapidly in the open air.
   (*c*) Hydrated seeds can only be stored at low temperature for a limited period.
   (*d*) Hydrated synthetic seeds should be planted soon after they are produced.

These problems can be resolved by providing a waxy coating over the beads. These problems are not associated with desiccated synthetic seeds.

METHODS OF ARTIFICIAL SEED ENCAPSULATION

There are two methods for encapsulation of synthetic seeds, *viz.*, dropping method and molding method. These are explained as follows:

1. Dropping Method

In this method, somatic embryos are encapsulated by dipping in hydro-gel. Hydro-gel used may include sodium alginate, agar from see weeds, seed gums like guar gum, locust bean gum, *etc.* Suppose we take sodium alginate solution (1– 5 per cent) as hydro-gel. Somatic embryos are dipped in this solution for encapsulation. After 20-30 minutes the artificial seeds are removed, washed with water and used for planting.

In another method, a burette is filled with sodium alginate solution (1– 5 per cent), dripped into a calcium nitrate solution (100 mM) drop by drop. Somatic embryo is inserted into the drop formed at the burette tip. Sodium alginate drop along with somatic embryos falls into the solution of calcium nitrate. Some chemicals like growth regulators, herbicides, insecticides, fungicides and mycorrhizae can be supplied to the somatic embryos during encapsulation along with the matrix. This method is applicable for embryo/auxiliary/apical/adventitious buds.

2. Molding Method

This method follows simple procedure of mixing somatic embryos with temperature dependent gel (*e.g.* gel rite, agar). The encapsulation of somatic embryos with the gel is achieved by lowering the temperature.

ADVANTAGES AND APPLICATIONS

Synthetic seed technology is useful for rapid easiest way of non-conventional method of propagation in crop plants. Some of the important advantages of synthetic seeds are given below:

1. **Rapid method:** In some plants it requires a long time to reach at reproductive phase and seed production. In these cases the synthetic seeds can be helpful to get the propagules in a short period. Generally the reproduction phase in a plant is season dependent, in these cases the somatic embryos and the synthetic seeds can be produced at any time as required.
2. **Long term storage:** In case of natural seeds there is a particular dormancy period, but in artificial seeds there is no dormancy period, thus it is more helpful for propagation. It can reduce the life cycle period of a plant.
3. **Somatic hybrids:** Artificial seeds can also be helpful where there is no successful seed production after sexual hybridization. The somatic embryos can be obtained from somatic hybrids obtained through protoplast fusion. Artificial seeds are helpful in case of meiotically unstable genotypes, where the normal seed set is of low frequency. In genetic manipulation of crop plants the production of artificial seed may be useful.
4. **Rapid embryo germination:** In cases where the embryo germination is difficult, the artificial seed can provide the beneficial adjuvants, *i.e.*, growth promoting seed substances, plant nutrients, *etc.* through the artificial coats.
5. **Maintains Variability:** Seeds produced artificially would produce same type of plants having similar life cycle. Harvesting and data management would be easy.
6. **Mass propagation:** In elite varieties were low frequency of normal seed set is observed, mass production of artificial seeds are important.
7. **Additional formation:** Synthetic seeds are produced by somatic cells and artificial embryo and seed coat formation. One can easily understand role of endosperm and seed coat formation by artificial seed production process. Somatic embryogenesis has been observed in a great many species to date, which indicates that it may be possible to produce artificial seeds in almost any desired crops Successful results have already been obtained in some crops such as *Apium graveolens Daucus carola, Zea mays, Lactuca satxva, Medicago sativa, Brassica sp. Gossypium hirsutum.*
8. **Germplasm conservation:** Recalcitrant seeds are defined as seeds which are sensitive to desiccation and cannot be dried below a certain critical moisture level. This level differs with different species, but is significantly higher than the 5-10 per cent moisture levels.
9. **Multiplication of Asexually Propagated Species:** Artificial seeds are successfully being applied in many of the non-seed producing plants, ornamental hybrids.

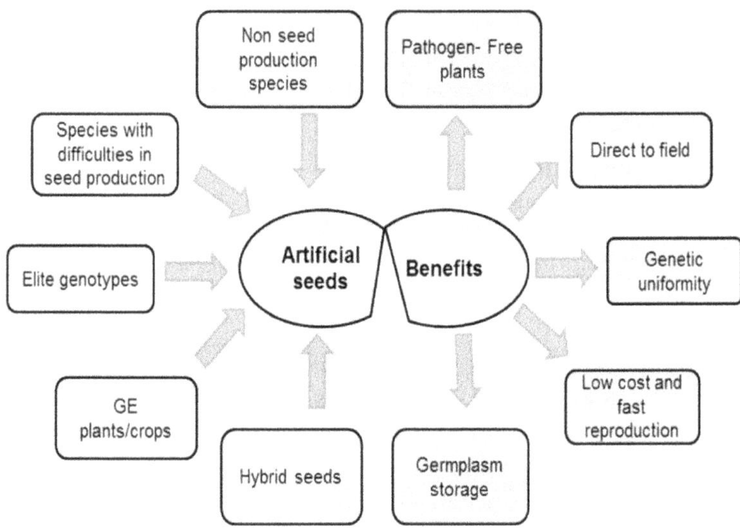

Some of the successful examples of artificial seeds

- Sujata singh (1997): Synthetic seed production in hybrid rice.
- Jayanthi (1999): Synthetic seed production in sweet potato.
- Neelamathi (2000): Synthetic seed production in sugarcane.
- Vakeswaran (2001): Synthetic seed production in Ashwagandha.
- Lakshmi (2001): Synthetic seed production in cassava.
- Nyende *et al.*(2008): Yield and canopy development of field grown potato plants derived from synthetic seeds.

LIMITATIONS

The main requirement for an efficient artificial seed production protocol is the large-scale production of highly valuable micopropagules suitable for encapsulation in sodium alginate matrix, at low cost per culture unit. However, although the design of such systems was achieved in some plant species, the micropropagation system is still one of the major limitations of the development of artificial seed technology.

1. Limited production of viable micropropagules that are useful in synthetic seed production.
2. Asynchrous development of somatic embryos.
3. Improper maturation of somatic embryos that makes them inefficient for germination and conversion into normal plants.
4. Lack of dormancy and stress tolerance in somatic embryos that limit the storage of synthetic seeds.
5. Somaclonal variations which may alter the genetic constituent of the embryos.

In some of the horticultural crops seeds propagation is not successful due to heterozygosity of seeds particularly in cross pollinated crops, minute seed size, *e.g.*, orchids, presence of reduced endosperm. Some seeds require mycorrhizal fungi association for germination, *e.g.*, orchids so no seeds are formed.

SUMMARY

Artificial seeds are artificially encapsulated somatic embryos (usually) or other vegetative parts such as shoot buds, cell aggregates, auxiliary buds, or any other micro-propagules which can be sown as a seed and converted into a plant under *in vitro* or in vivo conditions.

Important steps involved in the production of synthetic seeds include: selection of explants, induction of callus, induction of somatic embryos, proliferation of embryos, maturation of embryos, desiccation and tolerance induction, encapsulation and *in vitro* germination or transplantation in the field for germination.

There are two methods for encapsulation of synthetic seeds, *viz.*, dropping method and molding method. These are briefly explained. There are two methods for encapsulation of synthetic seeds, *viz.*, dropping method and molding method. These are explained. Synthetic seed technology is useful for rapid easiest way of non-conventional method of propagation in crop plants. Some of the important advantages of synthetic are discussed. An improved artificial seed production technique is considered a valuable alternate technology of propagation in many commercially important crops and a significant method for mass propagation of elite plant genotypes. However, despite the advantages of artificial seeds, further research is required in order to improve root formation of non-embryogenic artificial seeds.

GLOSSARY

Explant Material: The basic generative component of the artificial seed.

Somatic Embryos: Somatic embryos are the most common micropropagule used for artificial seed production.

Synthetic seeds: An encapsulated single somatic embryo.

QUESTIONS

1. **Define artificial seed and discuss its application in crop improvement.**
2. **Explain briefly advantages of synthetic seed.**
3. **What are major limitations in artificial seed production?**
4. **Describe different steps involved in artificial seed production.**
5. **Define desiccated and hydrated artificial seeds.**

6. **Write short notes on the following:**
   (*a*) Hydrated seed system (*b*) Desiccated seed system
7. **Give comparison of synthetic seeds and true seeds.**

*****

Chapter 15

Somatic Hybridization

INTRODUCTION

Somatic hybridization refers to crossing of crop plants through fusion of somatic cells. The fusion of somatic cells takes place through protoplasts. Naked cells or cells without cell wall are called protoplasts. The term protoplast was first used by Hanstein in 1880. Main points related to somatic hybridization are listed below:

1. It permits hybridization between any two plant species whether they are related or unrelated. Thus it overcomes the barriers of cross incompatibility.
2. In this method, there is equal contribution of cytoplasm from both parents, whereas in sexual hybridization cytoplasm is strictly contributed by maternal parent.
3. Somatic hybridization between two diploid species leads to production of allotetraploid hybrid, whereas sexual fusion between two diploid species leads to production of diploid hybrid.
4. It involves somatic cells (protoplasts) rather than sexual gametes bypassing sexual process.
5. Tissue culture is technique essential for regeneration of somatic hybrids into whole plants

TYPES OF SOMATIC HYBRIDS

Somatic hybrids can be obtained between sexually compatible as well as incompatible species. Somatic hybrids can be classified on the basis of taxonomic relationship and chromosome complements received from the parental species. Based on taxonomic relationship of species involved in the hybridization, somatic hybrids are of three types, *viz.*, interspecific hybrids, inter-generic hybrids and intertribal hybrids. These are briefly discussed as follows:

1. **Interspecific Hybrids:** Hybrids between two different species of the same genus are referred to as interspecific hybrids. Such hybrids have been obtained in tobacco, potato, carrot, Datura, Petunia, *etc.* Somatic interspecific hybrids have been obtained in sexually compatible as well as incompatible species of tobacco and Petunia and between incompatible species of *Datura*.
2. **Inter-generic Hybrids:** Hybrids between plants of two different genera of the same family are known as inter-generic hybrids. Somatic inter-generic hybrids have been obtained between tobacco + tomato, potato + tomato, Datura + *Atropa belladonna*, Arabidopsis + Brassica and several other plant species.
3. **Intertribal Hybrids:** Hybrids between plants of two different tribes are called intertribal hybrids. Such hybrids can only be produced by somatic hybridization. Such crosses exhibit more asymmetrical hybrids than interspecific and inter-generic hybrids.

1. Symmetric Hybrids

Those somatic hybrids which contain full chromosome complements of both the species involved in the fusion are referred to as symmetrical hybrids. In other words, such hybrids retain the somatic chromosome complements of both the fusion parents. **Such hybrids may** give rise to a new species.The well-known example is pomato. It has been obtained by the fusion of potato and tomato mesophyll protoplast. Such hybrids have been developed in tobacco, carrot, Petunia, *etc.* Some of the symmetric hybrids may be superior to their parents in some traits of economic value, and could be ultimately developed into useful crops. However, they often serve as useful sources of valuable genes. Many somatic hybrids have been produced between sexually incompatible species. Some of these hybrids possess and express useful genes, and are fertile. Symmetric hybrids provide the following opportunities in crop improvement programs.

(i) Production of hybrids between non-flowering or sterile lines.
(ii) Widening of the genetic base of an allopolyploid species.
(iii) Creation of a superior somatic hybrid.
(iv) Gene transfer from related species into cultivated species.
(v) Generation of novel materials for scientific studies.
(vi) Symmetric hybrids can also be used in a backcross program for transfer of desirable cytoplasm.

2. Asymmetric Hybrids

Those somatic hybrids which contain full chromosome complements of one species and only a part of somatic complement of another species involved in the fusion are referred to as asymmetrical hybrids. Such somatic hybrids exhibit the full somatic complement of one parental species few chromosomes of other parental species. Such hybrids are likely to show a limited introgression of chromosome segments from the eliminated genome due to drastically enhanced chromosome aberrations and/or mitotic crossing over *in vitro*. Asymmetric hybrids are produced due to spontaneous chromosome elimination in certain fusion combinations. Asymmetric hybrids are essentially cytoplasmic hybrids or cybrids except for the introgressed genes or chromosomes. Such hybrids have been obtained in tobacco, potato, Petunia and several other inter-generic and intertribal hybrids.

3. Cytoplasmic Hybrid or Cybrid

Those somatic hybrids which involve normal protoplast of one species and nucleusless (cytoplast) of another species are referred to as cybrids. In other words, cybrids are produced by fusion of protoplast of one parent and cytoplast of another parent. Cybrids posses nuclear genes from only one parent and cytoplasm from both parents. The objective of cybrid production is to be combine cytoplasmic genes of one species with nuclear and cytoplasmic genes of another species. Cybrids may be produced by fusion of a normal protoplast of one species with an enucleated protoplast (cytoplast) (or) a protoplast having an inactivated nucleus of other species. This provides an unique opportunity for transfer of plasmagenes from one species into nuclear background of another species in a single generation. The cybrid approach has been used for the transfer of cytoplasmic male sterility from *N. tabacum* to *N. sylvestris* from *Petunia* hybrid to *Petunia axillaries, etc.* and even in sexually incompatible combination.

STEPS INVOLVED

Somatic hybridization consists of five important steps, *viz.*, (1) isolation of protoplasts, (2) fusion of protoplasts, (3) selection of hybrid cells, (4) regeneration of plants from hybrid cells, and (5) culture of hybrid cells and characterization. These are briefly discussed as follows:

1. Isolation of Protoplasts

Protoplasts can be isolated by two methods, *viz.*, mechanical and enzymatic methods.

(a) Mechanical Method

This method is rarely used, because by this method protoplasts are obtained in small quantity and cells are broken. However, this method eliminates adverse effects of enzymes on the protoplasts. This method is suitable for isolation of protoplast from highly vacuolated cells of storage tissues, such as onion bulb, scales, radish roots, *etc.* However, this method has certain limitations which are as follows:

(i) It isolates small quantity of protoplasts.

(ii) It is a tedious and time consuming method.

(iii) It is not suitable for isolating protoplast from meristematic tissues and matured and less vacuolated cells.

(iv) The viability of protoplast is low because of the presence of substances released by damaged cell.

(b) Enzymatic Method

The possibility of enzymatic isolation of protoplast from higher plants was first demonstrated by Cocking in 1960. He used concentrated solution of cellulase to degrade the cell walls. In 1968, **Takebe *et al.*** for the first time employed commercial enzyme preparation for isolation of protoplasts and in 1971 regenerated first plant from protoplasts. By using this method protoplast can be isolated from any part of plant but it is easier to isolate protoplast from the mesophyll cells of leaf and pollen mother cells or tetrads. Young cell suspensions are particularly ideal for isolation of protoplasts in large quantities. The enzymatic method of protoplasts isolation can be performed in two ways as follows:

(i) Two Step Method

In this method, the tissue is first treated with macerozyme or pectinase enzyme which separates the cell by degrading the middle lamella. These free cells are then treated with cellulase enzyme which releases the protoplasts, cellulase enzyme digest the cellulose in plant cell walls while pectinase enzyme separates the pectin holding cells.

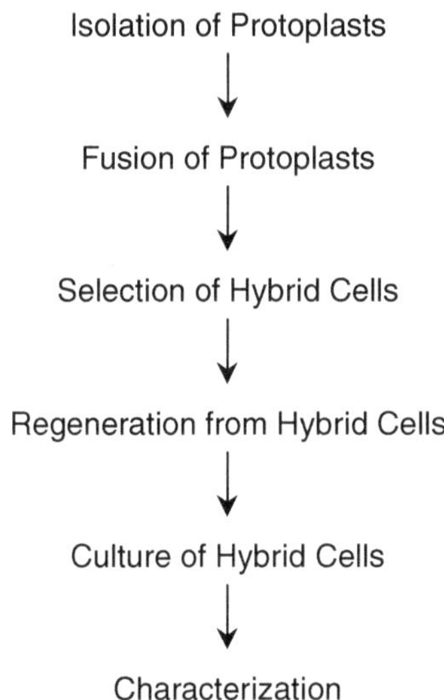

FIGURE 15.1: Steps Involved in Somatic Hybridization.

(ii) One Step Method

In this method, the tissue is subjected to a mixture of both macerozyme and cellulase enzymes in one step. The one step method is generally used because it is less labor intensive, while the yield of protoplast isolated by two step method are better for culture studies.

Advantages

There are several advantages of enzymatic method of protoplast isolation which are listed as follows:

(i) It permits isolation of large quantity of protoplast.
(ii) It minimizes deleterious effects of excessive plasmolysis.
(iii) The cells remain intact and are not injured as observed in mechanical method.

The main drawback of enzymatic method is that enzymatic digested mixture would contain sub-cellular debris, undigested cells, broken protoplast and healthy protoplast. This mixture is purified by the filtration, centrifugation and washing. The isolated protoplasts should be healthy and viable for rapid regeneration.

Sometimes, both methods can be combined for isolation of protoplasts to retain merits of both techniques. In combined method, first cells are separated mechanically and then used for isolation of protoplast through enzymatic reaction. Removal of cell wall makes possible fusion of the protoplasts of diverse origin and also uptake of foreign DNA.

Protoplasts can be isolated from any part of the plant, *viz.*, leaves, roots, pollen grains, pollen mother cells, callus culture, *etc.* Generally, protoplasts are isolated from mesophyll tissues or from callus cell suspension. Crop plants in which protoplasts can be isolated from mesophyll cells and callus cells are given below.

(i) Mesophyll Cells: Tobacco, *Datura*, Cucumber, Rapeseed, *Petunia*, Field pea, French bean, Cow pea, *etc.*
(ii) Callus Cells: Carrot, Sugarcane, Barley, Rice, Wheat, Soybean, Upland Cotton, *etc.*

The isolation of protoplasts from mesophyll cells involves four main steps, *viz.*, sterilisation of leaves, peeling off the epidermis, enzymatic treatment, and isolation and cleaning of protoplasts.

2. Protoplast Fusion

Two things are essential for protoplast fusion technology. Firstly protoplasts must be isolated in large quantity and secondly the isolated protoplasts should have the ability to regenerate in to new plants. The fusion of protoplasts involves following three main steps:

(i) **Agglutination:** During agglutination, the plasma membrane of protoplasts comes in close contact with each other.
(ii) **Membrane fusion:** After agglutination, membrane fusion occurs at localized places. This results in the formation of cytoplasmic channel or bridge. After fusion, mixing of the cytoplasm of two cells occurs. There is equal cytoplasmic contribution from both parents. The culture requirements of protoplasts are similar to those of single cells.

(iii) **Rounding off of the fused protoplast:** The fused protoplasts begin to become round due to expansion of cytoplasmic bridges thus forming spherical heterokaryons or homokaryons.

Two methods are used for protoplast fusion, *viz.*, chemical method and electrical method. These are explained as follows:

(i) **Chemical Method:** This method involves treatment of protoplasts with polyethylene glycol (PEG). This induces agglutination of the protoplasts and their fusion occurs after dilution of PEG with a solution containing high concentration of calcium ions at high pH. The frequency of fusion varies from 1 to 20 per cent depending upon cell type and fusion conditions employed.

(ii) **Electrical Method:** In this method, electric current is passed in the culture medium to induce fusion. This method leads to high frequency of protoplast fusion than chemical method. Moreover, it has less harmful effects on protoplasts than chemical method.

3. Selection of Hybrid Cells

After protoplast fusion, there is a mixture of parental types, homokaryons and heterokaryons. In other words, hybrid cells are of two types, *viz.*, homokaryons and heterokaryons. These are defined as follows:

(i) **Homokaryons:** The hybrid cells involving fusion of the protoplasts of the same species are called homokaryons. There will be homokaryons for both the species involved in protoplast fusion. Homokaryons are not the desired fusion products.

(ii) **Heterokaryons:** The hybrid cells involving fusion of the protoplasts of the two different species are called heterokaryons. These are the desired products of somatic hybridization. This portion represents less than 10 per cent of the total fusion products. Various methods are used for identification of heterokaryons or desired fusion products. The following three methods are commonly used for identification of desired fusion products.

(a) **Morphological Method:** Hybrid cells can be identified by hybrid vigour at the callus stage by morphology which is often intermediate between the parents.

(b) **Biochemical Method:** It includes use of isozyme analysis.

(c) **Molecular Technique**: It includes use of RFLP (restriction fragment length polymorphism) and Polymerase Chain Reaction (PCR) techniques for identification of hybrid cells.

4. Culture of Hybrid Cells and Characterization

Hybrid cells thus selected are cultured in a nutrient medium which is highly enriched with organic compounds. Growth hormones such as auxins and cytokinins are always required in the culture medium. Murashige and Skoog medium with suitable modification is used for culturing hybrid cells. In culture medium, protoplasts regenerate a new cell wall within 2-4 days and cell division starts within 2-7 days depending upon the plant species. Thus plants are regenerated from hybrid cells. Regenerated hybrid plants are characterized for highly heritable characters such as shape, size and color of leaf, stem, flower, *etc.* and stem surface (hairy or smooth).

APPLICATIONS

There are several potential applications of somatic hybridization in plant breeding such as gene transfer across species, development of somatic hybrids, conservation of heterosis, development of transgenic plants, production of fertile plants, modification of cytoplasmic genes, enhancement of photosynthetic efficiency, production of allotetraploid plants and hybrids in asexually propagated species. These are briefly presented as follows:

1. **Gene Transfer across Species:** Somatic hybridization overcomes barriers of cross incompatibility and permits gene transfer between distantly related species and unrelated species. It is a powerful tool for modification and improvement of polygenic traits. Genes for resistance to diseases, insects, drought, soil salinity, heat, cold, frost, lodging: quality improvement and male sterility can be easily transferred from related species or unrelated species. Disease resistance has been transferred in for leaf roll in potato from wild species of potato through protoplast fusion. Similarly, somatic hybrids produced between *Brassica napus* and *Eruca sativa* were fertile and had low oncentration of Erucic acid. Male sterile lines were developed by fusing protoplast of *N. tabacum* with X-ray irradiated protoplasm of *N. Africana.*

   Hybrids between two sexually incompatable species posses some desirable features that may make them commercially useful. For example, somatic hybrids between *Datura innoxa* x *D. discolor* and *D. innoxa x D. stramonium* show heterosis for alkaloid content. The alkaloid content is 20-25 per cent higher than their parents.

2. **Development of Somatic Hybrids:** Protoplast fusion permits development of somatic hybrids in different crop plants. It permits production of novel interspecific and intergeneric crosses between the plants that are difficult (or) impossible to hybridise conventionally. For example, pomato which is the result of cross between tomato and potato, though the combination is useless.

3. **Conservation of Heterosis**: In such hybrids, heterosis can be easily conserved.

4. **Development of Transgenic Plants:** Somatic hybrids are developed through protoplast fusion. Genetic transformation of hybrid cells (heterokaryons) will lead to development of transgenic hybrids.

5. **Production of Fertile Plants:** Some plants such as haploid, triploid and aneuploids are sexually sterile. These plants can be made fertile through protoplast fusion.

6. **Modification of Cytoplasmic Genes:** In somatic hybridization, cytoplasm is equally contributed by both the parents. Hybrids contain a mixture of two fusion partners. Thus modification of chloroplast and mitochondrial DNA is possible by this method.
7. **Enhancement of Photosynthetic Efficiency:** Somatic hybridization may enhance photosynthetic efficiency of plants through the transfer of efficient foreign chloroplast into the plants having low photosynthetic efficiency.
8. **Production of Allotetraploid Plants:** Somatic hybridization between two diploid species leads to production of allotetraploid plants in single step.
9. **Hybrids in Asexually Propagated Species:** Somatic hybridization permits production of in those species that propagate only by vegetative means such as potato, and some other tuber crops.
10. Somatic hybridization is possible between plants that are still in Juvenile phase.

ADVANTAGES

1. Symmetric hybrids can be produced between species which cannot be hybridized sexually. These hybrids can be readily used in breeding programs for transfer of useful genes to crops (or) may be useful as new crop species.
2. Hybrids can be produced even between such strains which are completely sterile such as monoploids.
3. Cytoplasm transfers can be effected in one year, which may take 5-6 years by back cross method. Even where back cross is not applicable cytoplasm transfer can be made using this approach
4. Mitochondria of one species can be combined with chloroplast of another species. This may be very important in some cases, and is not achievable by sexual means even between some easily crossable species.
5. Recombinant organellar genomes, especially of mitochondria, are generated in somatic hybrids and cybrids.

LIMITATIONS

There are some limitations of somatic hybridization which are listed as follows:

1. Somatic hybridization is successful only in those species where regeneration of plants is possible from protoplasts, *i.e.*, protoplasts are totipotent.
2. Generally, inter-specific somatic hybrids are fertile, whereas inter-generic and intertribal somatic hybrids are sterile.
3. Inter-specific somatic hybrids often contain many undesirable traits of wild species. Elimination of such undesirable traits through backcross is difficult.

4. The technique of protoplast isolation, culture, fusion and regeneration are not available for many important crop species like many cereals and pulses.
5. In many cases, chromosome elimination occurs from somatic hybrids leading them asymmetric hybrids. Such hybrids may be useful but there is no control on chromosome elimination.
6. Many somatic hybrids show genetic instability which may be an inherent feature of some species combinations.
7. Many somatic hybrids either do not regenerate or often sterile and unstable. Such hybrids are useless for crop improvement.

PRACTICAL ACHIEVEMENTS

Somatic hybrids have been developed in four major families families of crop plants namely Brassicaceae, Fabaceae, Poaceae and Solanaceae. Interspecific hybrids have been obtained in Nicotiana, Brinjal, Potato, Tomato, Brassica Medicago, Soyabean, Sugarcane, Datura, Petunia, Sorghum, *etc.* Intergeneric hybrids have been obtained between Tomato x Potato, Tomato x Tobacco, Festuca x Lolium, sugarcane x Bajra, Wheat x Bajra, Brasica x Eruca, *etc.* Inter tribal hybrids have been developed between *Brassica* x *Arabdopsis and Brassica* x *lesquerella Fendleri*

Somatic hybridization is a potential method of breaking barriers of cross incompatibility among various interspecific and intergenetic hybrids. It permits transfer of genes between species and also between genera. This technique has been used for transfer of desirable genes from wild species to cultivated species and also between two genera and different tribes. Moreover, this method is also useful in transfer of cytoplasmic male sterility from one species to another. Generally, disease resistance, quality, adaptation, drought resistance have been transferred from wild species to cultivated species. Two new crops, *viz.*, pomato (potato x tomato) and *Raphano brassica* (*Raphanus sativus x Brassica oleracea*) have been developed by somatic hybridization [protoplast fusion]. However, both have combination of undesirable traits.

SUMMARY

Somatic hybridization refers to crossing of crop plants through fusion of somatic cells. The fusion of somatic cells takes place through protoplasts. Naked cells or cells without cell wall are called protoplasts. The term protoplast was first used by Hanstein in 1880.

Based on taxonomic relationship of species involved in the hybridization, somatic hybrids are of three types, *viz.*, interspecific hybrids, inter-generic hybrids and intertribal hybrids. These have been briefly discussed.

Somatic hybridization consists of five important steps, *viz.*, (1) isolation of protoplasts, (2) fusion of protoplasts, (3) selection of hybrid cells, (4) regeneration of plants from hybrid cells, and (5) culture of hybrid cells and characterization. These are briefly discussed.

There are several potential applications of somatic hybridization in plant breeding such as gene transfer across species, development of somatic hybrids, conservation of heterosis, development of transgenic plants, production of fertile plants, modification of cytoplasmic genes, enhancement of photosynthetic efficiency, production of allotetraploid plants and hybrids in asexually propagated species. These are briefly presented.

Somatic hybrids have been developed in four major families families of crop plants namely Brassicaceae, Fabaceae, Poaceae and Solanaceae. Advantages of somatic hybridization have been discussed.

GLOSSARY

Asymmetrical somatic hybrids: Somatic hybrids that contain complete somatic complement of one species and only part of the somatic complements of other species.

Cybrids: Somatic hybrids with normal protoplast of one species and nucleusless protoplast of other species.

Cytoplast: A protoplast either without nucleus or with inactive nucleus.

Heterokaryons: Hybrid cell involving protoplasts of two different species.

Homokaryons: Hybrid cell involving protoplasts of the same species.

Inter-generic somatic hybrids: Hybrids obtained by protoplast fusion between two different genera of the same family.

Inter-specific somatic hybrids: Hybrids obtained by protoplast fusion between two different species of the same genus.

Inter-tribal somatic hybrids: Hybrids obtained through protoplast fusion between plants of two different families.

Protoplast: Naked cells or cells without cell wall.

Somatic hybridization: Crossing of crop plants through fusion of somatic cells.

Somatic Hybrids: Hybrids obtained by somatic hybridization.

Symmetrical somatic hybrids: Somatic hybrids that contain all chromosomes of both the species involved in the fusion of protoplast.

QUESTIONS

1. **Describe briefly different types of somatic hybrids with suitable examples.**
2. **Define somatic hybridization and describe its applications in crop improvement.**
3. **Describe briefly various steps involved in somatic hybridization.**
4. **Explain briefly inter-specific somatic hybrids with suitable examples.**
5. **Describe briefly inter-generic somatic hybrids with suitable examples.**

6. **Explain in brief practical achievements of somatic hybridization with examples.**
7. **Discuss briefly advantages of somatic hybridization.**
8. **Give comparison of somatic hybridization and sexual hybridization.**
9. **Explain limitations of somatic hybridization.**

*****

Chapter 16

Techniques of Foreign Gene Transfer

INTRODUCTION

The process of transfer, integration and expression of transgene in the host cells is known as genetic transformation. Successful transfer of foreign gene is essential to develop transgenic plants to solve specific problem or to improve crop plants for a particular character.

METHODS OF FOREIGN GENE TRANSFER

There are several techniques of foreign transfer. Most commonly used methods include *Agrobacterium* method, gene gun method and electroporation method. Procedure, advantages and limitations of each method are briefly discussed as follows:

1. *Agrobacterium* Method

This method is also known as vector mediated gene transfer or biological method of gene transfer or indirect method of gene transfer. In this approach the transgene is combined with a vector which takes it to the target cells for integration.

The *Agrobacterium* is a naturally occurring gram-negative soil bacterium with two common species *A. tumifacience* and *A. rhizogenes*. These are known as natural gene engineers for their ability to transform plants. *A. tumifacience* induces tumors called crown galls, whereas *A. rhizogenes* causes hairy root diseases. Large plasmids in these bacteria are called tumor inducing (*Ti* plasmids) and root inducing (*Ri* plasmid) respectively. This method is used with cells or protoplasts. It consists of the following four main steps.

(i) Gene Cloning

Gene cloning is a technique of genetic engineering which is used to make multiple identical copies of a gene. Plasmids of *Agrobacterium* contain tumour producing genes [Ti genes or oncogenes]. The foreign gene which has to be used for genetic transformation is incorporated in the plasmid for cloning. Thus several identical copies of the transgene are produced by self-replication of plasmids.

(ii) Genetic Transformation

The process of transfer, integration and expression of transgene or foreign gene in the host cells is known as genetic transformation. For genetic transformation, bacteria with cloned gene are mixed [in millions] in the cell culture or protoplast culture of the host plant for a day or two. Out of several thousands of cells or protoplasts in the culture medium, only a few are transformed. In other words, plasmid DNA [foreign DNA] can enter, integrate and express in few cells or protoplasts.

(iii) Identification of Transformed Cells

Several methods are available for detecting the presence of transformed cells. Such methods include selectable marker, PCR, ELISA, southern, northern and western blot techniques. A selectable marker is widely used for this purpose. The selectable marker is attached to the target gene for easy identification. Generally an antibiotic resistant gene is used as selectable marker to confirm whether the gene of interest has been inserted in the host plant or not. The Kanamycin resistant gene is used for identification of transformed cells. The lethal concentration of Kanamycin is put in the plasmid mixed cell suspension or protoplast culture. The non-transformed cells will die and only transformed cells will survive. Thus transformed cells are detected.

(iv) Regeneration of Transformed Cells

The transformed cells are transferred in to culture medium for transformation in to whole plant. The regeneration of cell or protoplasts in to whole plant is the basic requirement of this method of foreign gene transfer.

Advantages

There are two main advantages of *Agrobacterium* mediated method of DNA transfer. As given below.

(i) This method has some control over the copy number and site of integration of transgene which is not possible in particle bombardment method.

(ii) This is a cheaper method of genetic transformation.

Limitations

There are three main limitations of *Agrobacterium* mediated method of genetic transformation as given below.

(i) Host Specificity: This method has limited host range. It cannot infest monocots especially cereals.

(ii) Somaclonal variation: Lots of variation is induced in the tissue culture which poses problem in the identification of transformed cells.

(iii) Slow regeneration: Regeneration of cells or protoplasts into whole plant takes more time than regeneration from meristems and organ cultures.

2. Gene Gun Method

This method is known by various names such as particle bombardment method, micro-projectile method, biolistic method and particle acceleration method. In this method, foreign DNA is delivered into host plant cells using high velocity metal particles through gene gun. Main features of this method are given as follows:

(i) In this method regenerable tissues such as meristems, immature embryos and embryogenic callus are used for bombardment.

(ii) The gold or tungsten particles [1-1.5 milli mcrom size] are used for bombardment.

(iii) The gold or tungsten particle coated with gene of interest are bombarded in the tissues with the help of gene gun.

(iv) The transformed tissues are identified by polymerase chain reaction [PCR] technique.

(v) The transformed tissues are regenerated in culture medium into whole plant, which further examined for stability of gene expression by ELISA, southern blot and northern blot techniques.

Advantages

This method has some advantages over the *Agrobacterium* mediated method of DNA transfer. Advantages of this method are as follows:

(i) This method does not exhibit host specificity. Hence this method can be effectively used for development of transgenic plants in various plant species.

(ii) This method is technically simple than *Agrobacterium* mediated method of DNA transfer.

(iii) There is no need of isolating protoplasts. Various plant parts such as meristems, embryos or leaves can be bombarded and transformants can be regenerated from them.

TABLE 16.1: Comparison of Plasmid and Gene Gun Methods of Foreign Gene Transfer

Sl.No.	*Particulars*	*Plasmid Method*	*Gene gun Method*
1	Other term used	Biological method of gene transfer	Particle bombardment, microballistic and particle acceleration.
2	Basic requirement	Regenerable cells or protoplasts	Regenerable tissues or organs or cells.
3	Host specificity	Severe	No specificity
4	Effectiveness	More with dicots than monocots	Equally effective with both groups
5	Expenditure involved	Lower than gene gun method	Higher than plasmid method
6	Procedure	Difficult	Simple

Disadvantages

There are two main disadvantages of this method as stated below.

(i) There is no control over the copy number and site of integration of foreign gene.

(ii) The cost of equipment is high which prohibit use of this method by many researcher for DNA transfer.

3. Electroporaton Method

This method involves electroporation or chemical fusagens such as polyethyl glycol [PEG] with calcium phosphate. This method is used with protoplasts. A suspension of protoplasts with desired DNA is prepared. Then high voltage current is applied though the protoplast DNA suspension for a brief period. The electric current leads to formation of small temporary holes in the protoplast membrane though which the DNA can pass. After entry into the cell, the foreign DNA gets incorporated with the host genome, resulting in genetic transformation. These protoplasts are then cultured to regenerate in to whole plants. This method can be used in those crop species in which regeneration from protoplast is possible. Main points related to this method are given below:

(a) It is used with protoplast culture.

(b) This is cheaper method of foreign transfer.

(c) This is tedious than particle bombardment approach.

(d) It exhibit host specificity.

Currently two DNA delivery systems, *viz.*, (i) *Agrobacterium* mediated gene transfer, and (ii) particle bombardment are widely used for genetic transformation in cotton. In India, six different events have been developed by these methods, which are being used for developing *Bt.* transgenic cotton hybrids. In India, 522 *Bt.* cotton hybrids have been developed for commercial cultivation till 2009.

SUMMARY

The process of transfer, integration and expression of transgene in the host cells is known as genetic transformation. Most commonly used methods include agrobacterium method, gene gun method and electroporation method. Procedure, advantages and limitations of each method are briefly discussed.

Currently two DNA delivery systems, *viz.*, (i) *Agrobacterium* mediated gene transfer, and (ii) particle bombardment are widely used for genetic transformation in cotton. In India, six different events have been developed by these methods, which are being used for developing *Bt.* transgenic cotton hybrids. In India, 522 *Bt.* cotton hybrids have been developed for commercial cultivation till 2009.

The important steps in the development of transgenic plants include identification of the gene of interest, its isolation, cloning and transfer to the host plant.

GLOSSARY

Direct Gene Transfer: Introduction of foreign DNA into plant cells without the involvement of biological agents such as *Agrobacterium*.

Gene Gun Method: Delivery of foreign DNA into host plant cells using high velocity metal particles through gene gun.

Genetic transformation: The process of transfer, integration and expression of transgene in the host cells.

Indirect Gene Transfer: The vector mediated gene transfer or biological method of gene transfer or *Agrobacterium* mediated gene transfer.

QUESTIONS

1. **List methods of foreign transfer. Describe any one of them in detail.**
2. **Describe procedure, limitations and advantages of biological method of foreign gene transfer.**
3. **Discuss briefly procedure, limitations and advantages of particle method of foreign gene transfer.**
4. **Explain briefly procedure, limitations and advantages of micro-injection method of foreign transfer.**
5. **Describe briefly procedure, limitations and advantages of direct method of foreign gene transfer.**
6. **What are methods of gene transfer? Describe horizontal method of gene transfer. Give its merits and demerits.**
7. **Describe briefly vertical method of gene transfer. Give its merits and demerits.**
8. **Explain briefly differences between horizontal and vertical methods of gene transfer.**
9. **Write short notes on the following:**
   (i) Vertical gene transfer
   (ii) Horizontal gene transfer
   (iii) Particle bombardment method of gene transfer
   (iv) Biological method of gene transfer.

*****

Chapter 17

Restriction Enzymes and Cloning Vectors

INTRODUCTION

The restriction enzyme is a protein produced by bacteria and other prokaryotes that cleaves the DNA at specific sites. This site is known as the restriction site. They recognize and cleave at the restriction sites of the bacteriophage and destroy its DNA, to protect bacteria. Restriction enzymes are important tools for genetic engineering. They can be isolated from the bacteria and used in the laboratories. Each restriction enzyme recognizes specific one or a few restriction sites. When it finds its target sequence, a restriction enzyme will make a double-stranded cut in the DNA molecule. Over 3,000 restriction enzymes have been studied in detail, and more than 600 of these are available commercially.

TYPES OF RESTRICTION ENZYMES

Restriction endonucleases have two properties that are useful for recombinant DNA technology. Firstly, they cut the DNA fragments of size suitable for cloning and secondly, they make staggered cuts that create single stranded sticky end convenient to the formation of r-DNA.

Based on type of sequence recognized and nature of cut made in DNA, these enzymes have been classified into three different types, *viz.*, type I, type II, type and III. Recently lot more types have been recognized. Brief description of these types is presented as follows:

Type I Enzymes

Enzymes of this group are complex that cut DNA at random far from their recognition sequences. They are common.

These restriction enzymes cut the DNA far from the recognition sequences. However, they do not produce discrete restriction fragments, hence, are of not much practical value. *These are complex, multi-subunit restriction and modification enzymes. They were initially thought to be rare, but through genomic analysis they are found to be common and are of considerable biochemical interest.* Type I enzymes are of considerable biochemical interest, but they have little practical value since they do not produce discrete restriction fragments or distinct gel-banding patterns.

Type II Enzymes

These enzymes cut at defined positions closer to or within the restriction sites. Discrete restriction fragments and gel banding patterns are observed. They are exclusively used for DNA analysis and gene cloning in the laboratories. They are named after the bacterial species from which they are isolated, such as EcoRI is isolated from bacterial species *E. coli*. The restriction enzymes generate two different types of cuts. Blunt ends are produced when they cut the DNA at the centre of the recognition sequence, and sticky ends produce an overhang.

EcoRI Site

When EcoRI recognizes 5′-.GAATTC.-3′ 3′-.CTTAAG.-5′ and cuts this site, it always does so in a very specific pattern that produces ends with single-stranded DNA "overhangs":

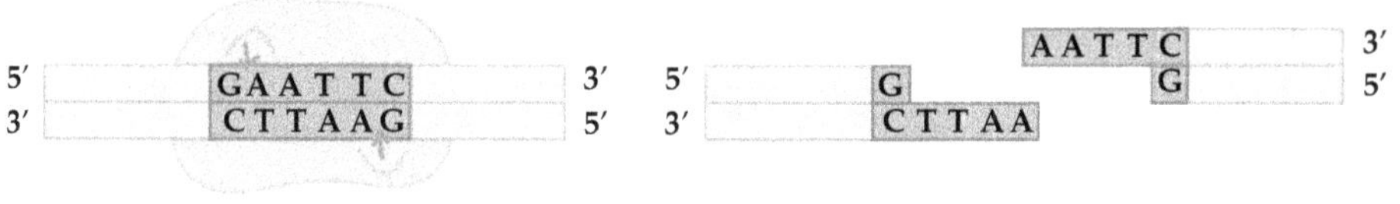

Ecori enzyme

If another piece of DNA has matching overhangs (for instance, because it has also been cut by EcoRI), the overhangs can stick together by complementary base pairing. For this reason, enzymes that leave single-stranded overhangs are said to produce **sticky ends**. Sticky ends are helpful in cloning because they hold two pieces of DNA together so they can be linked by DNA ligase. Not all restriction enzymes produce sticky ends. Some are "blunt cutters," which cut straight down the middle of a target sequence and leave no overhang. The restriction enzyme *Sma*I is an example of a blunt cutter:

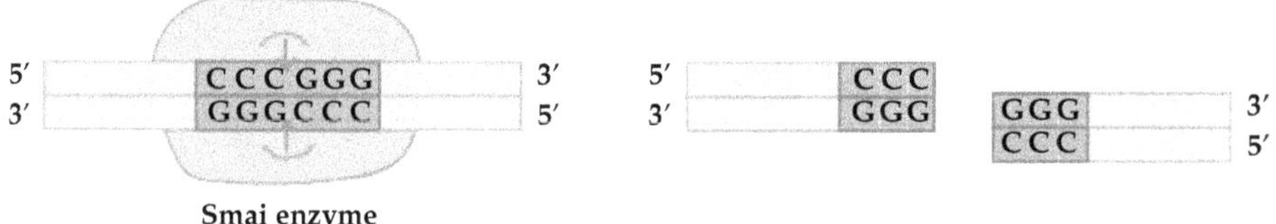

Smai enzyme

Blunt-ended fragments can be joined to each other by DNA ligase. However, blunt-ended fragments are harder to ligate together (the ligation reaction is less efficient and more likely to fail) because there are not single-stranded overhangs to hold the DNA molecules in position.

Type III Enzymes

Enzymes of this group are also large combination restriction-and-modification enzymes. They cleave outside of their recognition sequences and require two such sequences in opposite orientations within the same DNA molecule to accomplish cleavage; they rarely give complete digests.

DNA Ligases

The enzymes which are used to join DNA fragments are called DNA ligases. The DNA ligase enzyme is used to join the recombinant DNA fragment. Ligase requires 31 OH and 51 phosphate group for ligation. It joins the DNA fragment (or) seals the nicks between adjacent nucleotides in double stranded DNA reduces. There are two types of ligases, *viz.*, *E. coli* DNA ligase and T_4 bacteriophage DNA ligase. The T_4 enzymes requires ATP whereas the *E. coli* enzyme requires NAD+ both of these catalyst, the phosphodiester bond formation between the adjacent 31-OH and 5 – PO_4 termini in DNA. In the absence of DNA ligases r-DNA technique is not possible. The DNA ligases are used for the following purposes.

(i) Joining DNA fragments to produce the r-DNA molecule.

(ii) Ligation of vector and inserting the r-DNA.

(iii) Ligation of linkers/adopter/molecule at the blunt ends of fragments.

(iv) Sealing cuts in double stranded DNA.

APPLICATIONS OF RESTRICTION ENZYMES

There are several advantages of restriction enzymes. The important advantages include protection from viruses, generation of DNA fragments, synthesis of r-DNA, development of DNA markers, *etc.* These are discussed as follows:

1. **Protection from Viruses:** A restriction enzyme is a protein produced by bacteria that cleaves DNA at specific sites. Bacteria use restriction enzymes to defend against bacterial viruses called bacteriophages (or phage). When a phage infects a bacteria, it inserts its DNA into the bacteria so that it might be replicated. The restriction enzyme prevents replication of the phage DNA by cutting it into many pieces
2. **Generation of DNA Fragments:** DNA fragments are required for synthesis of r-DNA, generation of DNA markers, physical gene mapping. This can be achieved by using specific enzymes for cutting the DNA (restriction enzymes) into suitable fragments and then for joining together the appropriate fragments by DNA ligase.
3. **Synthesis of Recombinant DNA:** A recombinant DNA molecule is produced by joining together two or more DNA segments usually originating from different organisms. A recombinant DNA molecule is a vector into which the desired DNA fragment has been inserted to enable

its cloning in an appropriate host. The r-DNA molecules are produced with the following three objectives:

(i) To obtain a large number of copies of specific DNA fragments (achieved by genecloning)

(ii) To recover large quantities of the protein produced by the concerned gene.

(iii) To integrate the gene in question into the chromosome of a target organism where it expresses itself

4. **Development of DNA Markers:** The DNA markers especially restriction fragment length polymorphisms are generated with the help of restriction enzymes.
5. **RNAses are** used to remove the m-RNA from RNA-DNA hybrid in c-DNA cloning. It is also used to detect the presence of RNA – DNA hybrids and to remove poly A tails of m-RNA.
6. **The DNA ligases** are used for joining DNA fragments to produce the r-DNA molecule, ligation of vector and inserting the r-DNA, ligation of linkers and adopter molecule at the blunt ends of fragments and sealing cuts in double stranded DNA.
7. **Restriction** enzymes are used for construction of physical restriction maps, sequencing of large genomes, constructing gene libraries and molecular cloning.

CLONING VECTORS

DNA molecules which are capable of self-replication within a host cell and into which DNA sequences can be inserted and thus amplified are known as genetic vectors or cloning vector. Thus vector is a DNA molecule used to carry a gene of interest from one organism to another.

PROPERTIES OF A GOOD VECTOR

There are some qualities which are required in a good vector. Such properties include self-replication, isolation, easy for transfer, selectable markers, easy for identification of transformed cells, unique target sites and regulatory elements. These are explained as follows:

1. **Self-Replication:** The vector should be able to replicate independently of the replication of host chromosome. In other words, it should replicate autonomously.
2. **Isolation:** The isolation and purification techniques of the vector should be easy.
3. **Transfer:** Vectors are introduced in to the host cell for gene cloning. Hence, vector should be easily introduced into the host cells.
4. **Suitable Marker:** The vector should have suitable marker genes that allow easy detection or and selection of the transformed the host cell. It may include genes for resistance to antibiotics such as Ampicillin and Tetracycline.

5. **Easy Identification of Transformed Cells:** The identification of cells transformed with recombination DNA should be easy or selectable from those transformed by the unaltered vector.
6. **Unique Target Sites:** A vector should contain unique target sites for as many restriction enzymes as possible into which the DNA insert can be integrated.
7. **Regulatory Elements:** When expression of the DNA insert is desired, the vector should contain suitable regulatory elements such as promoter, operator and ribosome binding sites.

Types of Cloning Vectors

The most commonly used vectors include plasmid vectors, phagemid vectors, cosmid vectors, bacteria artificial chromosome (BAC) and yeast artificial chromosome (YAC). There are two commonly used vectors: plasmids and virus-based vectors. A detailed account of each type of cloning vector is presented as follows:

Plasmid

The size of a plasmid is 4361 bp, and the cloning limit is 0.1-10 kb. The marker gene is ampicillin and the tetracycline resistant gene whereas it remains isolated from the *E. coli*. Example: PBR322, p= plasmid, B= Bolivar (name of the scientist), R= Rodriguez (name of the scientist), 322= number of plasmid discovered in the same lab.

1. It is isolated from *E. coli*
2. Size: 4361 bp
3. Cloning limit: 0.1-10 kb
4. Marker gene: Ampicillin and Tetracycline resistant gene
5. Restriction site for various restriction endonucleases

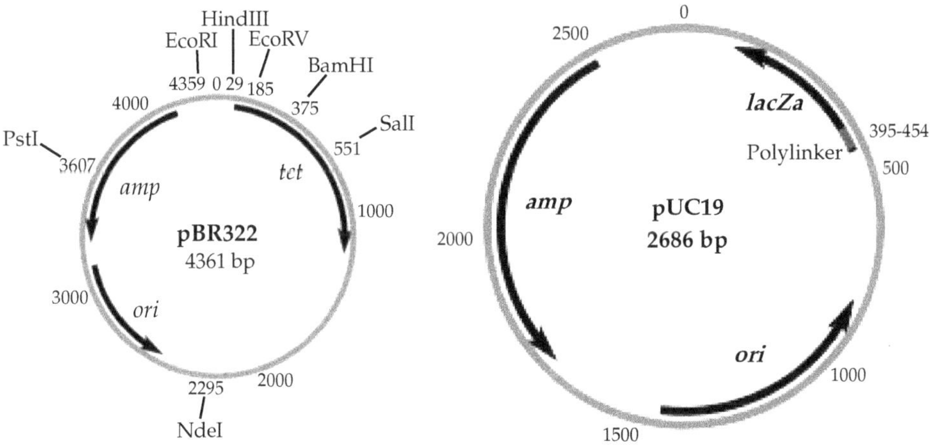

FIGURE 17.1: Plasmid Vectors.

pUC19 is one of a series of plasmid cloning vectors most widely used vector molecule for gene introgression. It can be easily distinguished from the non-recombinants based on colour differences of colonies on growth media. The designation "pUC" is derived from the classical "p" prefix (denoting "plasmid") and the abbreviation for the University of California, where early work on the plasmid series had been conducted. It is a circular double stranded DNA and has 2686 base pairs. pUC18 is similar to pUC19, but the MCS region is reversed.

Advantages of using Plasmids as Vectors

1. Easy to manipulate and isolate because of small size.
2. More stable because of circular configuration.
3. Replicate independent of the host.
4. High copy number.
5. Detection easy because of antibiotic-resistant genes.

Disadvantages of using Plasmids as Vectors

1. Large fragments cannot be cloned.
2. Size range is only 0 to 10kb.
3. Standard methods of transformation are inefficient.

BACTERIAL ARTIFICIAL CHROMOSOME

The size of BAC is 11827 bp and the cloning limit is 35-300 kb. The marker gene is chloramphenicol and lactose metabolizing gene. It is a modification of f-plasmid and is artificially synthesized. Example: pUvBBAC.

1. It is artificially synthesized plasmid.
2. size: 11827 bp.
3. It is modification of bacterial F-plasmid.
4. Cloning limit: 35-300 kb.
5. Marker gene: chloramphenicol resistant gene and lactose metabolizing gene (*LacZ*).

Advantages of BACs

1. They are capable of accommodating large sequences without any risk of rearrangement.
2. BACs are frequently used for studies of genetic or infectious disorders.
3. High yield of DNA clones is obtained.

Disadvantages of BACs

1. They are present in low copy number.
2. The eukaryotic DNA inserts with repetitive sequences are structurally unstable in BACs often resulting in deletion or rearrangement.

YEAST ARTIFICIAL CHROMOSOME

The size of YAC is 11400 bp and the cloning vector 100-1000 kb. It is artificial and has yeast centromere, which is isolated from the *Saccharomyces cerevisiae*. Example: pYAC3

1. It is an artificial chromosome having yeast centromere isolated from *Saccharomyces cerevisiae* and ligated to bacterial plasmid.
2. Size: 11400 bp.
3. It has telomere sequence.
4. Marker: similar as for identification of yeast cell.
5. Cloning limit: 100-1000 kb.

Advantages of using YACs

1. A large amount of DNA can be cloned.
2. Physical maps of large genomes like the human genome can be constructed.

Disadvantages of using YACs

1. Overall transformation efficiency is low.
2. The yield of cloned DNA is also low.

BACTERIOPHASE

Its size is 48502 bp and 1/3rd of this is not essential. It can only recombinant about 4-5 kbp of the donor DNA. An example is lambda genome.

Examples: Phage Lambda, M13 Phage, *etc.*

Phage Lambda I

1. It has head, tail, and tail fibers.
2. Its genome consists of 48.5 kb of DNA and 12 bp ss DNA which comprise of sticky ends at both the terminals. Since these ends are complementary, they are cohesive and also referred to as cos sites.
3. Infection by dikudd phage requires adsorption of tail fibers on the cell surface, contraction of the tail, and injection of the DNA inside the cell.

M13 Phage

1. These vectors are used for obtaining single-stranded copies of the cloned DNA.
2. They are utilized in DNA sequencing and *in vitro* mutagenesis.
3. M13 phages are derived from filamentous bacteriophage M13. The genome of M13 is 6.4 kb.
4. DNA inserts of large sizes can be cloned.
5. From the double-stranded inserts, pure single-stranded DNA copies are obtained.

Advantages of Phage Vectors

They are way more efficient than plasmids for cloning large inserts.

Screening of phage plaques is much easier than identification of recombinant bacterial colonies.

PHSAMID VECTORS

A phagemid vector is also known as phasmid vector. It is a type of cloning vector which is developed as a hybrid of the filamentous phage M13 and plasmids to produce a vector that can grow as a plasmid, and also be packaged as single stranded DNA in viral particles. Main points related to phagemid vectors are listed below:

1. A phagemid vector is also known as phasmid vector.
2. It is hybrid vector between bacteriophage and plasmid.
3. Phagemids contain an origin of replication (ori) for double stranded replication, as well as an f1 ori to enable single stranded replication and packaging into phage particles.
4. Many commonly used plasmids contain an f1 ori and are thus phagemids.
5. Similarly to a plasmid, a phagemid can be used to clone DNA fragments and be introduced into a bacterial host by a range of techniques (transformation, electroporation).

Filamentous phage retards bacterial growth but, in contrast to lambda and T7 phage, do not cause lyses. Helper phage are usually engineered to package less efficiently than the phagemid so that the resultant phage particles contain predominantly phagemid DNA.

Functions

Prior to the development of cycle sequencing, phagemids were used to generate single stranded DNA template for sequencing purposes. Today phagemids are still useful for generating templates for site-directed mutagenesis.

Several bacteria phages are used as cloning vectors, the most commonly used bacteria phage vectors are (lambda) phage, M-13 phages. These are explained as follows:

(i) Bacteriophage Vectors (Lambda)

There are three advantages of bacteriophage lambda vectors. They are more efficient than plasmids for cloning of large DNA fragments. Large size DNA fragment up to 25 kb can be cloned as compared to plasmid vector which can take up 10kb. Moreover, large number of phage plaques can be easily screened.

(ii) Phage M-13 Vector

They are derived from the 6.4 kb genome of the *E. coli* filamentous bacteriophage M13. This phage has a single stranded linear DNA genome in phase particles which converts into double stranded circular DNA molecule in the host cells. It contains origin of replication and a scorable marker gene *lac*-Z that compliments the gal host giving blue colonies. On transformation only white (or) clear plaques are obtained, thus permitting easy selection of recombinant plaques.

COSMIC VECTORS

Cosmids are extra-chromosomal circular DNA molecules that combine features of plasmids and phage. Their cloning limit is 35-50 kb. They were developed in the late 1970s and have been improved significantly since. The main features of cosmid vectors are as follows:

They are predominantly plasmids with a bacterial oriV, an antibiotic selection marker and a cloning site, but they carry one, or more recently two cos sites derived from bacteriophage lambda.

The loading capacity of cosmids varies depending on the size of the vector itself but usually lies around 40-45 kb.

Cosmids are essentially plasmids that contain a minimum of 250bp of DNA including *cos* site and sequences needed for binding of and cleavage by terminase.

A typical cosmid has origin of replication, selectable markers from a plasmid ampr and tetr and cloning sites (unique restriction sites) and Cos site (The sequence yielding cohensive ends). The typical features of cosmids are as follows:

1. Cosmids can be used to clone DNA inserts up to 45kb
2. Selection of recombinant DNA is based on the procedure applicable to the plasmid making up the cosmid.
3. Finally, these vectors are amplified and maintained in the same manner as the contributing plasmid such as cosmid PHC-79 and cosmid PJB-B.

Cosmid: The size of cosmid is 7900 bp and the cloning limit is 30-50 kb. It has features similar to both phase and plasmid and an example of it is super COS1. It has combined feature of both phase and plasmid

Advantages of using Cosmids as Vectors

1. They have high transformation efficiency and are capable of producing a large number of clones from a small quantity of DNA.
2. Also, they can carry up to 45 kb of insert compared to 25 kb carried by plasmids and dikudd.

Disadvantage

1. Cosmids cannot accept more than 50 kb of the insert.

HUMAN ARTIFICIAL CHROMOSOME

It is an artificial chromosome that is used to transfer human gene and has no limit on cloning as it can carry a large segment of the DNA.

Advantages of using HACs

1. No upper limit on DNA that can be cloned.
2. it avoids the possibility of insertional mutagenesis.

SHUTTLE VECTORS

These vectors are created by recombinant DNA technique. These are designed to replicate in cells of two different species. Since these vectors can be grown in one host and then moved into another without any extra manipulation they are called, shuttle vectors. Shuttle vectors have been designed to specifically satisfy the need *i.e.,* for the initial cloning of DNA inserts in *E. coli* and sub-sequent functional tests in the species to which the DNA inserts belong. Most of the Eukaryotic vectors are in facts shuttle vectors. Shuttle vectors contain two origins of replication, one specific for each host species, as well as those genes necessary for their replication. Some of them can be grown in two different prokaryotic species, while others can propagate in prokaryotic species (*E. coli*) and a eukaryotic one such as yeast, plants and animals.

STEPS INVOLVED IN GENE CLONING

It involves separating genes from the donor cell, inserting it to a small carrier molecule (Vector) and then replicating it into the host cell. Following basic steps involved in cloning:

Isolating the donor DNA gene: There are two ways in which the DNA fragment can be isolated. The first method is to use restriction endonuclease enzyme as it has a specific restriction site for its functioning. The second method is to use reverse transcriptase enzyme.

Selecting specific vector: An isolated gene does not have the capability of replicating itself. Hence, before insertion in the host, a specific vector must be chosen. The main function of this vector is to help in the replication process. There are specific features of the cloning vector that have been discussed in this article.

Incorporating the donor gene in the vector: The vector is split by the same RE enzyme that has been used for isolating of the DNA gene. The mixture of the vector and donor gene is combined and in the presence of the DNA ligase, a recombinant vector is formed in the mixture.

Transforming the recombinant vector to a host cell: The recombinant vector thus formed is then transformed into a host cell. Some cells are naturally transformable for example, *Bacillus*, while others are not naturally capable of transformation and hence necessitate the artificial method.

Isolating the recombinant cell: The host cell is then allowed to grow in culture media. However, this might contain both, recombinant and non-recombinant cells. This mixture needs to be separated for which the marker gene is used.

SUMMARY

An enzyme from bacteria that can recognize specific base sequences in DNA and cut the DNA at that site (the restriction site) is called restriction enzyme. Based on type of sequence recognized and nature of cut made in DNA, these enzymes have been classified into three different types, *viz.*, type I, type II, type and III. Recently lot more types have been recognized. Brief description of these types is presented.

There are several advantages of restriction enzymes. The important advantages include protection from viruses, generation of DNA fragments, synthesis of r-DNA, development of DNA markers, *etc.* These have been discussed.

A vector is a DNA molecule used to carry a gene of interest from one organism to another. The DNA molecule that carries foreign DNA into a host cell, replicates inside a bacterial (or yeast) cell and produces multiple copies of itself and the foreign DNA is called cloning vector.

There are some qualities which are required in a good vector. Such properties include self-replication, isolation, easy for transfer, selectable markers, easy for identification of transformed cells, unique target sites and regulatory elements. These have been explained.

The most commonly used vectors include plasmid vectors, phagemid vectors, cosmid vectors, bacteria artificial chromosome (BAC) and yeast artificial chromosome (YAC). There are two commonly used vectors: plasmids and virus-based vectors. A detailed account of each type of cloning vector is presented.

The important steps that are required for cloning of a vector include preparation of DNA and vector, union of DNA with the vector, transfer of DNA in to host cells and selection of transformed cells. These have been explained.

GLOSSARY

Bacterial artificial chromosome (BAC): An engineered DNA molecule which is used to clone DNA sequences in bacterial cells (for example, *E. coli*).

Cosmids: Extra-chromosomal circular DNA molecules that combine features of plasmids and phage.

Exonucleases: Enzymes which degrade nucleic acids starting at one or both the ends of polynucleotide chain.

Genetic Vector: A DNA molecule which is used to carry a gene of interest from one organism to another.

Ligase: Enzymes which is used to join DNA fragments.

Nucleases: Enzymes that degrade nucleic acids.

Phagemid: A type of cloning vector which is developed as a hybrid of the filamentous phage M13 and plasmids to produce a vector that can grow as a plasmid, and also be packaged as single stranded DNA in viral particles; also called phasmid.

Plasmid: DNA molecules other than bacterial chromosome that is capable of independent replication and transmission.

Restriction enzyme: An enzyme from bacteria that can recognize specific base sequences in DNA and cut the DNA at that site (the restriction site).

Reverse transcriptase: The enzyme which is used to synthesize c-DNA by using m RNA as template.

RNAses: Enzymes which are used to remove the m-RNA from RNA-DNA hybrid in c-DNA cloning.

Shuttle Vectors: These vectors are created by recombinant DNA technique to replicate in cells of two different species.

Terminal transferases : An enzymes that adds nucleotides to the 31 terminal of DNA molecule.

Yeast artificial chromosome (YAC): It is an artificial chromosome that contains telomeres, origin of replication, a yeast centromere, and a selectable marker for identification in yeast cells.

QUESTIONS

1. **Give a brief account of various types of restriction enzymes.**
2. **Give a brief account of DNA joining enzymes.**
3. **Explain briefly various advantages of restriction enzymes.**
4. **What is cloning vector? Describe briefly various properties of gene vector.**
5. **Explain briefly the properties of a good cloning vector.**
6. **List types of vectors. Describe any one of them in detail.**
7. **Describe briefly plasmid vectors on the basis of structure, function and types.**
8. **Give a brief account of phagemid and cosmid vectors.**
9. **Describe in detail bacterial artificial chromosomes as vectors.**
10. **Explain briefly yeast artificial chromosomes as vectors.**
11. **Explain briefly various steps involved in cloning of a vector.**
12. **Describe applications of vectors in plant biotechnology.**

*****

Chapter 18

Gene Isolation and Blotting Techniques

INTRODUCTION

The DNA isolation procedure makes use of the fact that lipids and proteins are soluble in chloroform and that the DNA is soluble in water (aqueous) solutions. In plants, the DNA isolation consists of six important steps, *viz.*, grinding of tissues, centrifugation of crushed tissues, adding chloroform, rotating of solution, transfer to tube and isolation of DNA. These are briefly discussed as follows:

1. **Grinding of Tissues:** Mechanically break-open the cell walls and membranes by grinding the tissue in a coffee grinder. The plant tissue should be frozen before grinding and ground in the coffee grinder with dry ice. Keeping the tissue frozen prevents enzymes in the cells from breaking down the DNA.
2. **Centrifugation of Pulverized Tissues:** Place the crushed tissue into a 50 ml centrifuge tube containing about 10 ml of CTAB buffer solution. The buffer solution contains compounds which disable the enzymes which will break down the DNA and also a compound called CTAB which will combine with the polysaccharides and proteins and chemically alter them. Mix the solution well to suspend the cellular material. Incubate the solution at 60°C for at least 30 minutes.
3. **Adding Chloroform:** Pour an equal volume (the same amount as is already in your tube) of chloroform into the 50 ml tube. "Extract" the solution by tilting or rocking the tube gently back and forth (if you mix the solution vigorously you may break your DNA molecules!). Chloroform is an organic compound and all of the cellular compounds which are soluble in chloroform (lipids, proteins) will be dissolved into the chloroform. DNA is not soluble in chloroform and will remain dissolved in the aqueous (water) layer.

4. **Rotation of Solution:** Rotate the solution in the centrifuge for 5 minutes. The gravitational forces in the centrifuge will completely separate the two phases according to density. Chloroform is very dense and will be the bottom layer, while the water layer is less dense and will be on the top.
5. **Transfer to Tube:** Transfer the supernatant (the aqueous layer containing the DNA) to a separate tube with a pipette.
6. **Isolation of DNA:** Add 2/3 volume of ice-cold isopropanol to the aqueous layer containing the DNA [supernatant]. The DNA dissolved in the supernatant will not be soluble in the ice-cold alcohol and will start to precipitate out of solution over a period of a few minutes. After a few minutes, gently rock the tube back and forth (if you rock or shake your tube hard you will break your DNA molecules). The DNA should appear as whitish strands to clear jelly-like strands which will float in the solution.

GENE CLONING

Gene cloning is a process by which multiple copies of a specific gene (DNA) can be made after isolation of such gene or DNA. In other words, gene cloning is the process of making several copies of a gene. Gene cloning is known by a variety of names such as DNA cloning, molecular cloning and recombinant DNA technology. In gene cloning, a segment of DNA representing a gene is inserted in to a cloning vector where it replicates further.

STEPS IN GENE CLONING

Gene cloning consists of five important steps, *viz.* isolation of donor DNA fragment or gene, selection of suitable vector, incorporation of donor DNA fragment into the vector, transformation of recombinant vector into a suitable host cell and isolation of recombinant host cell. These are briefly discussed as follows:

1. Isolation of Donor DNA Fragment or Gene

A donor DNA fragment should be isolated by suitable methods. There are two method for isolation of desired gene or DNA fragment. One is using restriction endonuclease enzyme: the enzyme restriction endonuclease is a key enzyme in molecular gene cloning. It has specific restriction site for its action. The enzyme RE generates a DNA fragment either with blunt end or with sticky end. Second is using reverse transcriptase enzyme: reverse transcriptase enzyme synthesizes complementary DNA strand of the desired gene using its mRNA.

2. Selection of Suitable Cloning Vector

When donor DNA fragment is incorporated into a host cell, it will not replicate because the isolated gene do not have the capacity to replicate itself. So before introduction of donor fragment into host, a suitable vector should be selected. Cloning vector is the DNA molecule capable of self-replication inside the host cell. The main function of cloning vector is to replicates the inserted DNA fragment inside the host cell. Some of the examples of cloning vectors are Plasmid, BAC, YAC,

DIKUDD-bacteriophage, expression vectors, *etc.* Basic Properties of cloning vector includes: 1. It must be self-replicating inside host cell, 2. It must possess restriction site for RE enzymes, 3. Introduction of donor DNA fragment must not interfere with replication property of the vector, 4. It must possess some marker gene such that it can be used for later identification of recombinant cell.

3. Incorporation of Donor DNA Fragment with Plasmid Vector

The plasmid vector is cut open by the same RE enzyme used for isolation of donor DNA fragment. The mixture of donor DNA fragment and plasmid vector are mixed together. In the presence of DNA ligase, base pairing of donor DNA fragment and plasmid vector occurs forming recombinant vector in the mixture.

4. Transformation of Recombinant Vector into Suitable Host

The recombinant vector is transformed into suitable host cell (bacterial cell). Some bacteria are naturally transformable because they take up the recombinant vector automatically such as *Bacillus, Haemophillus and Helicobacter pylori*. Some other bacteria are not naturally competent, in those bacteria recombinant vector are incorporated by artificial method such as Ca^{++} ion treatment, electroporation, *etc.*

5. Isolation of Recombinant Cell

The recombinant host cell is then grown in culture media but the culture may contains colonies both recombinant cell and non-recombinant cell. For isolation of recombinant cell from non-recombinant cell, marker gene of plasmid vector is employed. For examples, PBR322 plasmid vector contains different marker gene (Ampicillin resistant gene and Tetracycline resistant gene). When pst1 RE is used it knock out Ampicillin resistant gene from the plasmid, so that the recombinant cell become sensitive to Ampicillin.

Modern molecular cloning methods are used for production of recombinant proteins, vaccine production, gene therapy, antisense technology and genetic engineering in agricultural crops. It gives an amazingly potent toolbox for exploring, manipulating and strapping up DNA that will further widen the perception of science.

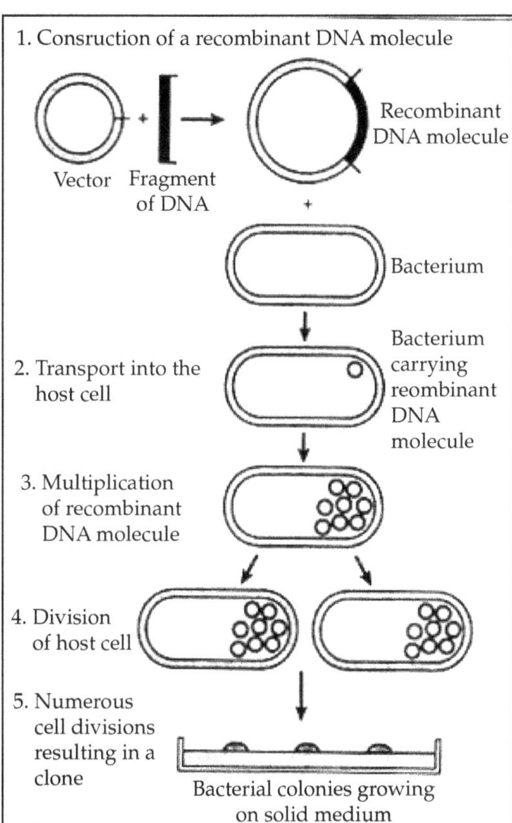

FIGURE 18.1 Steps in Gene Cloning.

Genetic Transformation of *E. coli*

DNA can be exchanged among bacteria by three methods: transformation, transduction and conjugation. Transformation is one of the most popular techniques of molecular genetics because it is often the best way to reintroduce experimentally altered DNA into cells. This technique was first discovered in bacteria, but other ways have been designed to transform many types of animal and plant cell as well. Transformation was the first mechanism of bacterial gene exchange to be discovered. The initial experiment on transformation was performed by Frederick Griffith in England in 1928 when he was working with two strains of *Streptococcus pneumoniae*, in 1944. Oswald T. Avery, Colin M. MacLeod, and Maclyn McCarty demonstrated that the "transforming principle" was DNA. Both results are milestones in the study of the molecular nature of genes. Two elements are required in a transformation system, suitable host bacterium and plasmid. For this, commonly *E. coli* is being used as host organism. Plasmid encodes some enzymes and antibiotic resistant markers which are expressed in the bacterium after transformation. Plasmids are small, circular DNA found naturally in many bacteria. Plasmid is found as an extra-chromosomal DNA and it contains some genes that the bacterium would not normally possess.

There are two forms of transformation: natural and artificial, each process depends on the ability of the organism to transform the DNA into the host cells. In natural transformation, bacteria are capable of transforming DNA naturally which means they can take up DNA from their environment directly. In artificial transformation, bacteria are not naturally transformable which they do not take up DNA from the environment. Bacterial cells have been exposed to particular chemical or electrical treatments to make them more permeable and then only the cells can take up DNA efficiently. Bacteria can take up DNA artificially by using different techniques such as electroporation, heat shock, Ca^{2+} treatment of cells and protoplast uptake of DNA. In these techniques, cells made permeable to DNA by calcium ion treatment will take up both single stranded and double stranded DNA. Therefore, both linear and double stranded circular plasmids can be efficiently introduced into chemically treated cells. This fact has made calcium ion induced competence very useful for cloning and other applications that require the introduction of plasmid and phage DNA into cell. The process of bacterial transformation and autonomous replication of the engineered plasmid DNA allows the production of large amounts of the DNA of interest within the bacterial host. This allows for further manipulations of the cloned DNA or for the expression of the gene of interest in the bacteria itself. The produced protein may result in a new bacterial phenotype, or the protein itself may be the desired end product.

Principle of Transformation is based on facts that transformation process allows a bacterium to take up genes from its surrounding environment. There are two major parameters involved in efficiently transforming a bacterial organism induce competence for transformation and second major parameter is the genetic constitution of the host strain of the organism being transformed. Competent cells are capable of up taking DNA from their environment and expressing DNA as functional proteins. Calcium chloride treatment is one of the best methods for the preparation of competent cells. Bacteria can take DNA from the environment in the form of plasmid. Most of them are double stranded circular DNA molecules and many can exist at very high copy numbers within a single bacterial cell. Blue-white selection (Alpha complementation) can be used to determine which plasmids carry an inserted fragment of DNA and which do not. These plasmids contain an additional gene (*lac* Z) that encodes for a portion of the enzyme β–galactosidase. When it transformed into an appropriate host, one containing the gene for the remaining portion of β–galactosidase, the intact enzyme can be produced and these bacteria form blue colonies in the presence of X–gal (5-bromo-4-chloro-3-indoyl-b-D-galactoside) and a gratuitous inducer called IPTG (Isopropyl β–D-Thiogalactopyronoside). These plasmids contains a number of cloning sites within the *lac* Z gene, and any insertion of foreign DNA into this region results in the loss of the ability to form active β–galactosidase. Therefore colonies that carry the plasmid with the insert, *i.e.*, Transformants will remain white and the colonies without the foreign DNA (Non-Transformants) will remain Blue.

GEL ELECTROPHORESIS

Gel electrophoresis is a technique used to separate DNA fragments, RNA or proteins based on their size and charge. Electrophoresis involves running a current through a gel containing the molecules of interest. Based on their size and charge, the molecules will move through the gel at different speeds, allowing them to be separated from one another. All DNA molecules have the same amount of charge per mass. Because of this, gel electrophoresis of DNA fragments separates them based on size only. Using electrophoresis, we can see how many different DNA fragments are present and how large they are relative to one another. We can also determine the absolute size of a piece of DNA by examining it next to a standard "yardstick" made up of DNA fragments of known sizes.

Agarose is isolated from the seaweed genera *Gelidium* and *Gracilaria*, and consists of repeated agarobiose (L- and D-galactose) subunits. When the agarose is heated in a buffer and allowed to cool, it will form a solid, slightly squishy gel. At the molecular level, the gel is a matrix of agarose molecules that are held together by hydrogen bonds and form tiny pores. At one end, the gel has pocket-like indentations called **wells**, which are where the DNA samples will be placed:

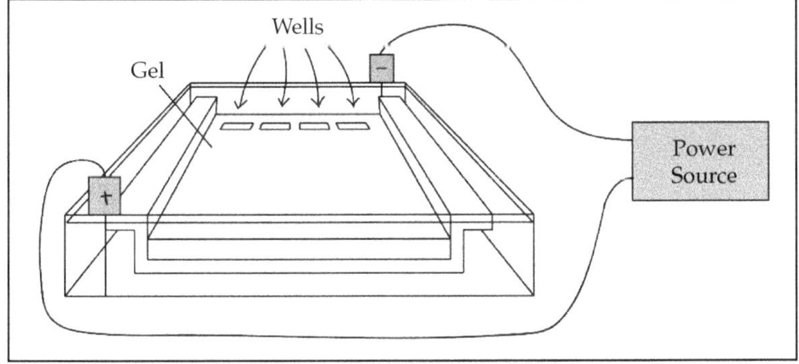

FIGURE 18.2: Gel Electrophoresis

One end of the electrophoresis tank is hooked to a positive electrode, while the other end is hooked to a negative electrode. The main body of the box, where the gel is placed, is filled with a salt-containing buffer solution that can conduct current. DNA samples are loaded into wells at negative electrode end of gel. Power is turned on and DNA fragments migrate through gel towards the positive electrode. After the gel has run, the fragments are separated by size. The largest fragments are near the top of the gel (negative electrode, where they began), and the smallest fragments are near the bottom (positive electrode). As the gel runs, shorter pieces of DNA will travel through the pores of the gel matrix faster than longer ones. Once the fragments have been separated, we can examine the gel and see what sizes of bands are found on it. When a gel is stained with a DNA-binding dye commonly ethidium bromide is being used, and placed under UV light, the DNA fragments will

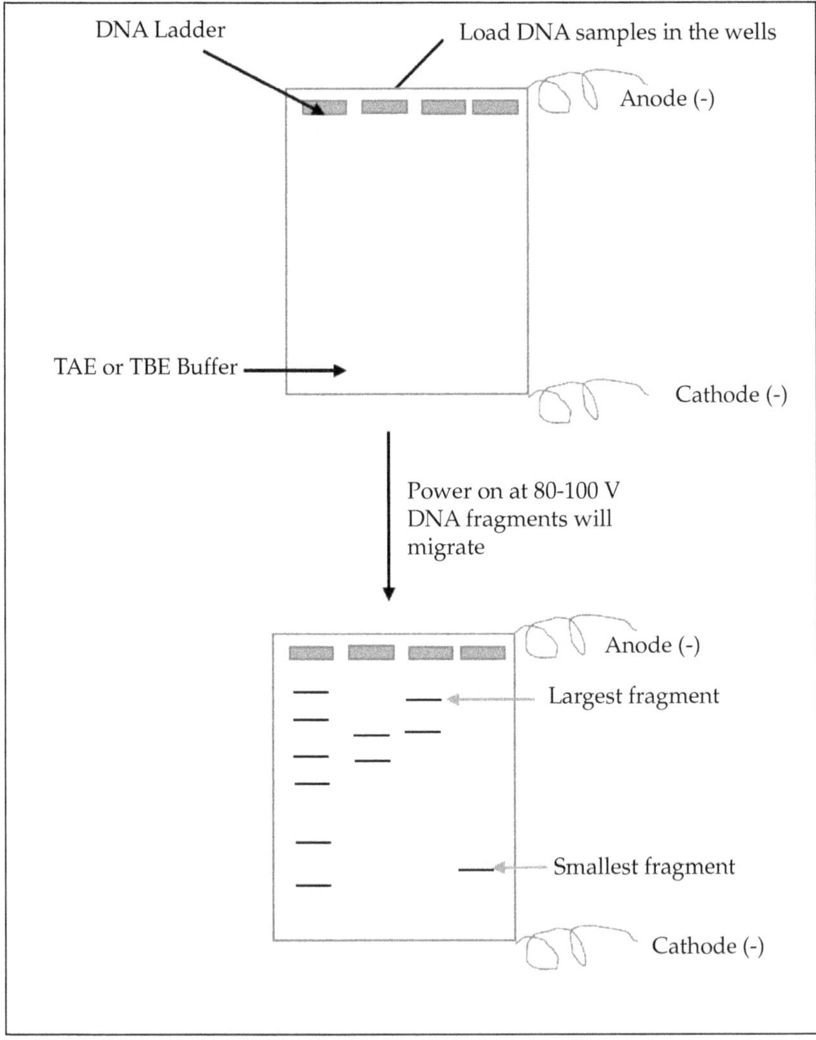

FIGURE 18.3: Fragments Separated by Size.

glow, allowing us to see the DNA present at different locations along the length of the gel. A well-defined "line" of DNA on a gel is called a **band**. Each band contains a large number of DNA fragments of the same size that have all travelled as a group to the same position. By comparing the bands in a sample to the DNA ladder, we can determine their approximate sizes.

PREPARATION OF PROBES

Preparation of probes for hybridization involves *in vitro* incorporation of reporter molecules into nucleic acids. These reporters can be incorporated at one or both ends of nucleic acid molecules, giving specific, low density labelled probes. High density labelling is usually achieved by incorporating the reporters uniformly throughout entire length of nucleic acid molecules. Two types of reporter molecules are presently in use, radioactive reporters and nonradioactive reporters. Radioactive reporters are usually tagged with 32P or 35S isotopes and can be directly detected. Non-radioactive reporters can be fluorescent tags, permitting their direct detection, or specific ligands such as biotin, haptens and hapten-like molecules, the detection of which is indirect. Labelling nucleic acids with radioisotopes, fluorophores, biotin, or digoxigenin enables their detection and analysis. When designing a labelling strategy, consider the intended application, the source of nucleic acid, and the type of label to incorporate. DNA oligonucleotides can be 52 endlabelled with radioisotopes in a reaction catalysed by T4 polynucleotide kinase, or nonisotopic labels can be incorporated into oligonucleotides during DNA synthesis. Larger DNA substrates can be labelled by 52 end labelling (radioisotopes) or labelled uniformly along the length of the DNA by nick translation or random primed synthesis (using radioisotope or non-isotopic labels). The labelled DNA can be used for a variety of applications, including probing Southern blots, probing northern blots, *in situ* hybridization, quantifying realtime PCR results, and gel shift assays.

BLOTTING TECHNIQUES

The presence of gene of interest in a sample can be detected by determining the presence of three components, *viz.*, DNA, mRNA and protein. There are three main techniques which are used to detect the presence of gene of interest in a sample. These are southern blotting, northern blotting and western blotting. All these techniques make use of gel electrophoresis.

Southern Blotting

Southern blotting is a hybridization technique for identification of particular size of **DNA** from the mixture of other similar molecules. This technique is based on the principle of separation of DNA fragments by gel electrophoresis and identified by labelled probe hybridization. Southern blotting was introduced by Edwin Southern in 1975 as a method to detect specific sequences of DNA in DNA samples. Southern blotting is an example of RFLP (restriction fragment length polymorphism). Basically, the DNA fragments are separated on the basis of size and charge during electrophoresis. After transferring separated DNA fragments on nylon membrane, the desired DNA is detected using specific DNA probe that is complementary to the desired DNA. A hybridization probe is a short (100-500bp), single stranded DNA. The probes are labelled with a marker so that they can be detected after hybridization.

Basic Procedure of southern blotting includes Restriction digestion by RE enzyme, Gel electrophoresis, Denaturation by Treating with HCl and NaOH, Blotting (transfer of bands from gel to nitrocellulose membrane), Baking and Blocking with casein in BSA, Hybridization using labelled probes and Visualization by autoradiogram.

1. **Restriction digestion:** The DNA is fragmentized by using suitable restriction enzyme.
2. E cuts the DNA at specific site generating fragments. The number of fragments of DNA obtained by restriction digest can be amplified by PCR
3. **Gel electrophoresis:** The desired DNA fragments is separated by gel electrophoresis
4. **Denaturation:** The SDS gel after electrophoresis is then soaked in alkali (NaOH) or acid (HCl) to denature the double stranded DNA fragments. DNA strands get separated
5. **Blotting:** The separated strand of DNA is then transferred to positively charged membrane nylon membrane (Nitrocellulose paper) by the process of blotting.
6. **Baking and blocking:** After the DNA of interest bound on the membrane, it is baked on autoclave to fix in the membrane. The membrane is then treated with casein or Bovine serum albumin (BSA) which saturates all the binding site of membrane.
7. **Hybridization with labelled probes:** The DNA bound to membrane is then treated with labelled probe. The labelled probe contains the complementary sequences to the gene of interest. The probe bind with complementary DNA on the membrane since all other non-specific binding site on the membrane has been blocked by BSA or casein.
8. **Visualization by Autoradiogram:** The membrane bound DNA labelled with probe can be visualized under autoradiogram which gives pattern of bands.

Applications

1. Southern blotting technique is used to detect DNA in given sample.
2. DNA fingerprinting.
3. Used for paternity testing, criminal identification, victim identification.
4. To isolate and identify desire gene of interest.
5. Used in restriction fragment length polymorphism.
6. To identify mutation or gene rearrangement in the sequence of DNA.
7. Used in diagnosis of disease caused by genetic defects.
8. Used to identify infectious agents.

Northern Blotting

A northern blot is a laboratory method used to detect specific RNA molecules among a mixture of RNA. Northern blotting can be used to analyse a sample of RNA from a particular tissue or cell type in order to measure the RNA expression of particular genes. This method was named for its similarity to the technique known as a Southern blot. The first step in a northern blot is to denature, or separate, the RNA within the sample into single strands, which ensures that the strands are unfolded and that there is no bonding between strands. The RNA molecules are then separated according to their sizes using a method called gel electrophoresis. Following separation, the RNA is transferred from the gel onto a blotting membrane. (Although this step is what gives the technique the name "northern blotting," the term is typically used to describe the entire procedure.) Once the transfer is complete, the blotting membrane carries all of the RNA bands originally on the gel. Next, the membrane is treated with a small piece of DNA or RNA called a probe, which has been designed to have a sequence that is complementary to a particular RNA sequence in the sample; this allows the probe to hybridize, or bind, to a specific RNA fragment on the membrane. In addition, the probe has a label, which is typically a radioactive atom or a fluorescent dye. Thus, following hybridization, the probe permits the RNA molecule of interest to be detected from among the many different RNA molecules on the membrane.

Western Blotting

Western blotting is an important technique used in cell and molecular biology. By using a western blot, researchers are able to identify specific proteins from a complex mixture of proteins extracted from cells. The technique uses three elements to accomplish this task: (1) separation by size, (2) transfer to a solid support, and (3) marking target protein using a proper primary and secondary antibody to visualize. Western blotting was first described by Harry Towbin in 1979. Western blotting (or immune-blotting) is a widely used method to detect proteins as well as post-translational modifications on proteins, using antibody based probes to obtain specific information about target proteins from complex samples. It is a routine method in a molecular biology, biochemistry and cell biology field with the multitude of applications. It can provide semi-quantitative or quantitative data about the target protein in simple or complex biological samples. Since western blotting is a multistep protocol, variations and errors can occur at any step reducing the reliability and reproducibility of this technique. Recent reports suggest that a few key steps, such as the sample preparation method, the amount and source of primary antibody used, as well as the normalization method utilized, are critical for reproducible western blot results. This method relies on the fact that most epitopes (sites recognized by antibodies, generally comprising several amino acids) in spite of denaturation of proteins can still be recognized. Due to high affinities of antibody toward their epitopes, it is a very sensitive method and even pico-gram quantities of a target protein can be detected. The two primary advantages of western blotting are sensitivity and specificity.

PCR TECHNIQUE AND ITS APPLICATION

Polymerase Chain Reaction (PCR) is a revolutionary method developed by Kary B. Mullis in the year 1983 and she was awarded by Nobel Prize for this discovery of chemistry in 1993. PCR is based on using the ability of DNA polymerase to synthesize new strand of DNA complementary to the template strand of DNA quickly and accurately. This technique is having an impact on many areas of molecular cloning, genetics, recombinant DNA research molecular biology, forensic analysis, evolutionary biology, and medical diagnostics. It is capable of taking a small amount of DNA or even single molecule and amplifying a specific region. Before the development of PCR, the methods used to amplify or generate copies of recombinant DNA fragments were time-consuming and labour-intensive. But PCR reactions can complete many rounds of replication and produce billions of copies of a DNA fragment only in few hours. The three major steps in a PCR cycling reactions, which are repeated up to 20 to 40 cycles denaturation, annealing and extension. It is always done on an automated thermos-cycler, which has ability to heat and cool the reaction tubes in a very short period of time. The Basics of PCR Cycling are given below:

(i) **Denaturation (94°C):** During this stage, the double strand melts open to form single stranded DNA, all enzymatic reactions stop.

(ii) **Annealing (54°C):** Hydrogen bonds are constantly formed and broken between the single stranded primer and the single stranded template. If the primers exactly fit the template, the hydrogen bonds formed are so strong that the primer stays attached.

(iii) **Extension (72°C):** The bases (complementary to the template) are coupled to the primer on the 3′ side (the polymerase adds dNTP's from 5′ to 3′side, reading the template from 3′ to 5′ side, bases are added complementary to the template).

Every cycle results in a doubling of the number of strands in the DNA present. After starting stage of few cycles, most of the product DNA strands made are the same length as the distance between the primers. The result is amplification of DNA that exists between the primers. The amount of amplification is 2 raised to the n power; n represents the number of cycles that are performed. After 20 cycles, this would give approximately 1 million fold amplification. After 40 cycles the amplification would be 1 x 1012.

APPLICATIONS OF PCR IN AGRICULTURE

The polymerase chain reaction is utilized by a wide range of researchers in a continually expanding scope of experimental orders.

1. PCR can be utilized for hereditary testing.
2. PCR can be utilized for Genetic fingerprinting.
3. PCR can be utilized for DNA fingerprinting can help in Parental testing (DNA sequencing).

4. PCR can be utilized for DNA cloning as it can concentrate fragments for insertion into a vector from a bigger genome, which may be accessible in little amounts.
5. PCR can be utilized for the investigation of examples of quality expression.
6. PCR can be utilized to build up connections among species.
7. PCR can be utilized in paleo-history.
8. PCR can be utilized to identify transgene.
9. For genetic purity testing.
10. For backcross breeding.
11. For gene isolation and identification.
12. For gene tagging and allele mining.

SUMMARY

The DNA isolation procedure makes use of the fact that lipids and proteins are soluble in chloroform and that the DNA is soluble in water (aqueous) solutions. In plants, the DNA isolation consists of six important steps, *viz.*, grinding of tissues, centrifugation of crushed tissues, adding chloroform, rotating of solution, transfer to tube and isolation of DNA. These have been discussed.

There are two principal ways of obtaining DNA fragments, *viz.* restriction endonuclease enzymes and Polymerase Chain Reaction. The applications of these techniques in obtaining DNA fragments have been discussed.

The presence of gene of interest in a sample can be detected by determining the presence of three components, *viz.*, DNA, mRNA and protein. There are three main techniques detect the presence of gene of interest in a sample. These are southern blotting, northern blotting and western blotting.

A small nucleotide sequence [15-30bp] of DNA or RNA which is used to detect the presence of complimentary sequences of nucleic acid samples is referred to as probe. Both the DNA and RNA sequences are used as probes which can be obtained from either genomic DNA or c-DNA libraries.

DNA probes are useful in identification of r-DNA clones, plant varieties, transgenes, criminals and paternity; isolation of DNA segment; development of RFLP maps and genomic maps of eukaryotes and disease diagnosis.

GLOSSARY

DNA: Deoxyribose nucleic acid; the genetic material.

DNA Isolation: The process of isolating [separating] DNA from plant and animal tissues.

DNA Probe: A small nucleotide sequence [15-30bp] of DNA or RNA which is used to detect the presence of complimentary sequences of nucleic acid in a sample.

Gene Detection: The process of detecting the presence of a particular gene in the DNA sample.

Northern Blotting: A technique used in molecular biology research to determine the presence of a specific mRNA from a set of different samples of RNA.

PCR: A technique widely used in molecular biology for gene amplification and DNA polymorphism.

Southern blotting: A method routinely used in molecular biology to detect the presence of a specific DNA sequence (a gene) in a DNA sample.

Western Blotting: A method of detecting the presence of a specific proteins in a given sample of tissues or extract using antibody specific to that protein.

QUESTIONS

1. **Describe briefly various steps involved in DNA isolation protocol.**
2. **Discuss methods of extraction of DNA fragments.**
3. **Describe the role of Polymerase Chain Reaction in isolation of DNA fragments.**
4. **How gel electrophoresis is useful in isolation of DNA fragments?**
5. **Write short notes on the following:**
   (i) Gel electrophoresis (ii) Polymerase Chain Reaction
6. **Define southern blotting. Explain various aspects of this technique.**
7. **What is northern blotting? Explain various aspects of this technique.**
8. **Define Western blotting. Describe various aspects of this technique.**
9. **What are gene detection techniques? Describe any one of them in detail.**
10. **Give a brief comparison of Southern blot, Northern blot and Western blot techniques.**
11. **Define the following terms:**
    (i) Southern blotting (ii) Northern blotting
    (iv) Western blotting (iv DNA probes

*****

Chapter 19

Transgenic Technology

INTRODUCTION

The technique or process of developing transgenic individuals is referred to as transgenic technology. It is also known as gene technology, transformation technology, genetic engineering and recombinant DNA technology. The crop plants which are developed through the application of plant biotechnology or plant genetic engineering are called transgenic plants or genetically modified plants or genetically engineered plants. In broad sense transgenic plants can be called as genetically modified plants. But in strict sense, they are genetically engineered plants. However, the term GM crops or GM plants or GM organisms has been very popular and is widely used the world over. The GM plants have been developed to solve specific problems which could not be solved easily through conventional plant breeding methods.

TRANSGENIC PLANTS

The genetically engineered plants are called transgenic plants. Main features of transgenic plants are listed as under.

1. Transgenic plants contain transgene or foreign gene of special significance.
2. Transgenic plants are developed through plant biotechnology.
3. In the development of transgenic plants, the sexual process is not involved.
4. The transgenic plants are recovered at a very low frequency.
5. Transgenic plants are developed to solve those problems which cannot be solved by conventional plant breeding techniques.

ADVANTAGES OF TRANSGENIC TECHNOLOGY

Transgenic technology has several advantages such as rapid method, free gene transfer, single gene transfer, direct gene transfer and solution to difficult problems. These are briefly discussed as follows:

1. **Rapid Method:** It is an effective and rapid method of crop improvement. It takes 3-4 years for developing new cultivars/hybrids against 10-15 years taken by conventional methods.
2. **Free Gene Transfer:** It permits gene transfer across the species, genera, family even from unrelated organisms.
3. **Single Gene Transfer:** Gene technology permits transfer of one or two genes from donor parent to the recipient parent. In conventional hybridization method, hundreds of genes are transferred to the recipient parent. Many of transferred genes are undesirable. Elimination of such genes requires repeated backcrossing to the recipient parent which takes 4-5 years.
4. **Direct Gene Transfer:** Gene technology permits direct gene transfer into the recipient parent bypassing sexual process. In other words, there is no need of union of male and female gametes in gene technology. The gene of interest can be directly inserted into the cell of recipient parent.
5. **Solution to Difficult Problems**: Gene technology provides solution to those problems that cannot be solved by conventional methods of breeding. The best example is resistance to bollworms in cotton.

APPLICATIONS OF TRANSGENIC TECHNOLOGY

Transgenic technology has several practical applications. Important applications of transgenic technology include improvement in (i) biotic resistance, (ii) abiotic resistance, (iii) herbicide resistance, (iv) quality of food products, and (v) development of novel traits such as golden rice and male sterility. All these aspects are briefly discussed as follows:

1. **Resistance to Biotic Stresses:** Biotic stress refers to adverse effects on crop growth and yield by biotic factors such as insects, diseases and parasitic weeds. In crop plants, heavy yield losses are caused every year due to insect and disease attack. Moreover, insecticides and pesticides which are used to control insects and diseases are expensive and have adverse effects on other beneficial organisms (parasites and predators). Gene technology has played key role in developing insect resistant cultivars in several crops such as bollworm resistant cultivars in cotton and stem borer resistant cultivars in maize. Moreover, the technology is eco-friendly.
2. **Resistance to Abiotic Stresses**: Abiotic stress refers to adverse effects on crop growth and yield by abiotic factors such as drought, soil salinity, soil acidity, cold, frost, *etc.* The cold resistant genotypes in tobacco and freezing resistant genotypes have been developed through gene technology. Efforts are being made to develop drought and salinity resistant cultivars in many crops.
3. **Herbicide Resistance:** In crop plants, weeds cause heavy yield losses and also adversely affect quality of the produce. The genetic resistance is the cheapest and the best way of solving this problem. Gene technology has been used to develop herbicide resistant cultivars in cotton, maize,

wheat, tobacco, potato, tomato, rapeseed, soybean, flax, *etc.* In these crops, cultivars resistant to glyphosate, gluphosinate and some other herbicides have been developed.

4. **Improvement in Quality:** The quality is adjudged in three ways, *viz.*, nutritional quality, market quality [keeping quality] and industrial quality. Gene technology has helped in improving these qualities in different crops. For example, the ripening and softening in tomato has been delayed. It is desirable for safe transport and storage. This has been achieved by manipulating the genes that encode the enzyme responsible for ripening [ethylene forming enzymes] and softening [polygalactonase]. Other quality improvement include non-browning potato, starch composition of wheat flour, carotene content in rice and improved oil content in oilseed crops.
5. **Development of Novel traits:** Plant biotechnology plays important role in development of novel traits. Examples are golden rice and transgenic male sterility in rapeseed.
6. **Industrial Products:** Gene technology has great potential for the production of biodegradable plastics, obtaining therapeutic proteins, pharmaceuticals and edible vaccines from transgenic plants. It may also help in producing biodiesel or petroleum products.
7. **Bioremediation:** It is possible to develop plants which can be used for bio-remediation of sick soils.

PRACTICAL ACHIEVEMENTS

Plant biotechnology has played significant role in (i) improving resistance [to insects, diseases, drought, salinity, frost, cold, *etc.*, (ii) food quality, and (iii) inducing male sterility in some crops. Some examples of such achievements are cited below [Table 19.2].

1. Improvement in Resistance

It includes insect resistance, disease resistance, salinity resistance and herbicide resistance. Such resistance has been achieved in several crops as mentioned below.

(i) **Insect Resistance**: It includes bollworm resistance in cotton, European borer resistance in maize and Colorado beetle resistance in Potato. Resistance to these insects has been transferred from *Bacillus thuringiensis.*

(ii) **Disease Resistance**: Resistance to Pierce disease in grapes has been developed at the University of Florida. Resistance to mosaic in papaya, potato and squash has been achieved in USA.

(iii) **Salinity Resistance**: Salinity resistant tomato has been developed at the University of California and at the University of Toronto.

(iv) **Herbicide Resistance**: It has been achieved in cotton, Flax, Soybean, Wheat, Potato, Tomato, Rapeseed, *etc.* The gene for herbicide resistance has been transferred from Steptomyces to wheat, potato and tomato; and from microbial gene in cotton, soybean and linseed.

2. Improvement in Quality

Quality improvement has been achieved in several crops. Some examples are cited below.

(i) **High Starch Potato**: In potato starch contents have been increased by transferring gene from human intestine bacteria [*Escherichia coli*].

(ii) **High Protein Potato and Alfalfa**: Protein contents have been improved in potato and alfalfa. In potato, the high protein gene has been transferred from amaranth and serum albumin gene from human. In alfalfa ova-albumin gene has been transferred from chicken.

(iii) **High Methionine Soybean**: In soybean, methionine content has been enhanced by transferring gene from a bacterium [*Bertholletia excelsa*].

(iv) **High Carotene Rice**: In rice, high carotene content gene has been transferred from Daffodils and **Golden rice** has been developed. The golden rice is rich in vitamin A content. Two genes have been transferred from Daffodils and one from bacterial species.

(v) **Anti-cracking Tomato**: In tomato, anti-cracking gene has been transferred from winter flounder fish. This has increased the transportation and keeping quality of tomato.

(vi) **Freezing Resistant Tobacco**: In tobacco, freezing resistant gene has been transferred from winter flounder fish and cold resistant gene from *Arabidopsis thaliana*.

3. Male Sterility

The male sterility has been developed in rapeseed through the application of agricultural biotechnology. The *barnase* and *barstar* system has been developed in this crop. The male sterility gene has been transferred from *Bacillus amyloliquefaciens*.

TABLE 19.1: Traits Improved in some Genetically Modified Crop Plants

Sl.No.	*Traits Improved*	*Examples of GM crops*
1	Insect Resistance	Cotton, Corn, Tobacco, Potato, *etc.*
2	Disease Resistance	Potato, Tobacco, Squash, Grapes, *etc.*
3	Freezing Resistance	Tomato, Tobacco, *etc.*
4	Salinity Resistance	Tomato
5	Herbicide resistance	Cotton, Corn, Soybean, Rapeseed, Alfalfa, Linseed, Sugar beet,
6	Quality	
i	Starch content	Potato
ii	Protein content	Potato, Alfalfa, *etc.*
iii	Methionine	Soybean
iv	Carotene content	Golden Rice
v	Oleic acid content	Soybean
7	Male sterility	Rapeseed

TRAITS IMPROVED AND GENE SOURCES USED

Several plant traits in GM crops have been improved through the application of plant biotechnology and different gene sources have been used for such improvement. Some examples are given in Table 19.2.

TABLE 19.2: Traits Improved and Gene Sources Used in some GM Crops

Sl.No.	*Crop*	*Traits Improved*	*Gene used From*
1	Soybean	Herbicide resistance Methionine	Microbial gene *Bertholletia excelsa*
2	Maize	Insect resistance [ECB] Herbicide resistance	*Bt.* Gene *Arabidopsis thaliana*
3	Cotton	Bollworm resistance Herbicide resistance	*Bt.* Gene Microbial gene
4	Canola	Male sterility Herbicide resistance	*Bacillus amylolifaciens* *Arabidopsis thaliana*
5	Papaya	Ring spot virus	Wild species
6	Squash	Mosaic virus	Coat protein gene
7	Potato	Protein content Mosaic resistance Herbicide resistance	Serum albumin gene of human and Amaranth Potato virus x and y Streptomyces
8	Tomato	Insect resistance Freezing resistance Mosaic resistance	*Bt.* Gene Winter Flounder Fish Tomato mosaic virus
9	Tobacco	Insect resistance Herbicide resistance Cold resistance Freezing resistance	Cowpea TI gene Acaligene Eutrophus *Arabidopsis thaliana* Winter Flounder Fish
10	Alfalfa	Protein quality	Chicken ova-albumin gene

PROBABLE RISKS OF PLANT BIOTECHNOLOGY

There are some probable risks of plant biotechnology such as adverse effects on human and animal health, development of resistance in some weeds against herbicides, ethical issues and development of resistance in insects and diseases against transgenic plants. These are explained as follows:

1. **Adverse Effects:** Consumption of transgenic food and vegetables may have adverse effects on human and animal health such as toxicity, allergenicity Six transgenic crops, *viz.* soybean, maize, cotton, potato and sugar beet are grown in Australia for human consumption. However, no adverse effects have been reported so far. The effects of GM crops on health

and environment are uncertain. The cultivation of such crops can have unintended adverse effects on both animal health and the environment.

2. **Resistance to Herbicides:** There is fear that continuous application of herbicides may lead to development of resistance in weeds against such herbicides. Development of such resistance has been reported in Johnson grass, Amaranth palmeri and many other weeds against glyphosate herbicide in USA.
3. **Ethical Issue:** The transfer of gene from animal such as from fish to potato and tobacco; ova-albumin gene of chicken to alfalfa and serum albumin gene of human into potato and plants will have ethical problems.
4. **Antibiotic Resistance**: Wide spread cultivation of transgenic plants may lead to development of resistance in insect pests and pathogens against transgenic plants.

FUTURE SCOPE

It is expected that in future plant biotechnology [GM crops] can revolutionize world agriculture, particularly in developing countries. This technology will substantially improve food security, reduce malnutrition, and increase rural income. In future, the transgenic research on development of GM crops needs to be directed towards the following thrust areas.

- **Improvement in Resistance:** There is need to develop GM crop cultivars resistant to drought and soil salinity in countries like India.
- **Nutritional Quality:** There is need to develop GM crop cultivars of fruits and vegetables with improved nutritional, market and keeping qualities.

  Crops to be Improved: It is expected that in future, the main focus would be on the development of GM cultivars in rice (stem borer), brinjal (fruit borer), potato (mosaic resistance), and wheat (herbicide resistance).

There is no need of developing herbicide resistant crop cultivars, because there is huge application of herbicides in weed control which pollutes the soil, underground water and has adverse effects on soil micro-organisms [earth worms, N_2 fixing bacteria, blue green algae, *etc.*]

In future, to achieve agricultural revolution through gene technology, the following points should be given due importance.

1 **Affordable Technology:** The new technology should be affordable by the farmers of developing-world. Moreover, the farmers must understand how to use them.

2 **Research by Public Sector:** There is a need for larger investments in research in the public sector. Partnerships between the public and private sectors can result in more efficient production of GM crops that are useful to the developing world and can expand the accessibility of those crops and their associated technologies to developing-world farmers.

3. **Due Importance to Agriculture:** Policymakers should give due importance to agriculture to fight hunger and malnutrition in near future particularly in under developed and developing countries.
4. **Proper Regulatory System:** Policymakers in the developing world must set regulatory standards that take into consideration the risks as well as the benefits of foods derived from GM crops. Proper regulatory system is essential to curb malpractices associated with seeds of GM crops.
5. **Specific Applications:** Plant biotechnology should be used only to solve those problems that cannot be addressed by conventional breeding techniques or to create novel genotypes for various purposes.

SUMMARY

Plant biotechnology has wide applications in the field of agriculture, medicines and industry. It opens countless doors of research in these areas. Biotech crops do not contain yield enhancing gene. The improvement in yield is achieved through effective control of insects, diseases and weeds. Biotech crops have been accepted by many developed and developing countries of the world. Plant biotechnology provides solution to certain difficult problems that cannot be solved by conventional breeding techniques. Examples are insect resistance [bollworm in cotton, European corn borer in maize, bud worm in tobacco, fruit worm in tomato and brinjal, *etc.*], resistance to mosaic disease [papaya, squash, potato, *etc.*] and herbicide resistance. It also provides solution to malnutrition and hunger. Despite these, biotech crops need to be used with precautions to avoid their unwarranted effects on non-target insects, ecosystem, non-biotech crops, human and animal health. Moreover, plant biotechnology is not a substitute for plant breeding. It can certainly be used as a tool to speed up the crop improvement programs.

GLOSSARY

***Agrobacterium tumifaciens*:** A soil-borne bacterium which is widely used for development of transgenic plants in different crop plants. It is a biological method of foreign gene transfer into host cells.

Callus: A mass of unorganized regenerated cells in culture medium.

Explant: The plant part which is used for regeneration into whole plant in tissue culture.

Gene Cloning: The process of making several identical copies of a gene in genetic engineering.

Genetic Engineering: Isolation, introduction and expression of foreign DNA in Plants and animals.

Genetic Transformation: The process of transfer, integration and expression of transgene (foreign gene) in an organism.

Micro-Injection: A mechanical method of gene transfer in which foreign DNA is delivered into the host cells by microscopic needles also known as micro-injection.

Particle Bombardment: A method of foreign gene transfer which involves high velocity metal particles (usually gold or tungsten) or metal particles mediated method of genetic transformation. It is also known as biolistic micro-projectile bombardment, micro-ballistics and particle acceleration.

Plant Biotechnology: Combined study of plant tissue culture and genetic engineering or recombinant DNA technology.

Plant Tissue Culture: Growth of tissues of living plants in a suitable culture medium (*invitro*). It includes cell culture, protoplast culture, organ culture and meristem culture.

Plasmid: An extra-chromosomal genetic element which is found in the bacteral cell, can replicate independently of chromosomal DNA and is not essential for normal growth and development of bacterium.

Protoplasts: Naked cells without cell wall.

Somaclonal variation: The variability generated by the use of tissue culture technique.

Transformants: Cells or tissues of an organism which have udergone genetic transformation; resultants of genetic transformation.

Transgenes: Foreign genes or modified genes of the same species which are used for the development of transgenic individuals. Such gene may be from the same species (in modified form), related wild species, unrelated species and microbes (bacteria, fungi and viruses).

Transgenic Breeding: Genetic improvement of crop plants, domestic animals and useful micro-organisms through biotechnology.

Transgenic: Genetically engineered organisms or organisms developed by the techniques of genetic engineering. It may be a plant, an animal or micro-organism.

QUESTIONS

1. **Define transgenic. Describe briefly various features of transgenic plants.**
2. **What is transgenic technology? Discuss its various advantages.**
3. **What are various methods of DNA delivery in crop plants? Describe any one of them.**
5. **Discuses briefly the procedure, advantages and limitations of *Agrobacterium* mediated system of DNA delivery in crop plants.**
6. **Explain the procedure, advantages and limitations of gene gun method of foreign gene transfer in crop plants.**
7. **Give a brief comparison of plasmid method and Particle bombardment method of foreign gene transfer in crop plants.**

8. **Describe with suitable examples the role of transgenic technology in genetic improvement of crop plants.**

9. **Give a brief account of important achievements of transgenic technology in crop plants.**

10. **Give a brief comparison of the following:**
    (*a*) Transgenic breeding and conventional breeding.
    (*b*) Plasmid method and particle bombardment method of DNA delivery in crop plants.

11. **Define the following terms:**
    (*a*) Transgenics (*b*) Plasmids
    (*c*) Transgenes (*d*) Transformants
    (*e*) Particle bombardment (*f*) Explant

*****

Chapter 20

Introductory Bioinformatics

INTRODUCTION

Advances in molecular biology and computer science over the past 30-40 years resulted in the development of bioinformatics. Bioinformatics is an interdisciplinary field of molecular biology and genetics relating with computer science, mathematics, and statistics. Term Bioinformatics was invented by Paulien Hogeweg and Ben Hesperin 1970. Bioinformatics basically develops methods and software tools to understand biological data. Word "In-silico" (To perform on computer) is strongly correlated with bioinformatics. The key areas of bioinformatics include biological databases, sequence alignment, gene and promoter prediction, molecular phylogenetic, structural and functional bioinformatics.

APPLICATIONS OF BIOINFORMATICS

Bioinformatics has several important applications in agriculture such as varietal identification, studies on omics, pedigree analysis, crop modelling, biometrical analysis, forecasting models, *etc.* these are briefly discussed as follows:

1. Varietal Identification and Screening

Bioinformatics is very useful in developing varietal information system. Bioinformatics is playing key role in DUS testing for varietal identification, grouping of varieties on the basis of various highly heritable characters, screening out of cultivars for use in Pre-breeding and traditional breeding, classification of PGR based on various criteria, *etc.* It helps in retrieval of data belonging to specific trait group such as early maturity, late maturity, dwarf types, tall types, resistant to biotic stresses/diseases, resistant to abiotic stresses/drought, submergence, salinity, genotypes with superior quality parameters, genotypes with specific novel genes, *etc.*

2. Studies on Omics

Basic set of chromosomes is known as genome. In a genome, each type of chromosome is represented only once. The study of structure and function of entire genome of an organism is referred to as genomics. The term genomics was coined by Thomas Roderick in 1986. Structural Genomics deals with the study of the structure of entire genome of an organism while Functional Genomics deals with the study of genome function, mainly deals with transcriptome and proteome. Transcriptome refers to complete set of RNAs transcribed from a genome and proteome refers to complete set of proteins encoded by a genome. Proteomics refers to the study of structures and functions of all proteins in an individual. Metabolomics refers to the study of all metabolic pathways in a living organism. Bioinformatics deals with all analysis of genomics, proteomics and metabolomics. Metabolomics is computer based information about metabolic pathways in a living organism. It helps in identification and correction of metabolic disorders in an organism by early detection of genetic disorders associated with metabolic pathways. It also helps in selection of individuals with normal metabolic pathways.

3. Pedigree Analysis

Computer aided studies are useful in pedigree analysis of various cultivars and hybrids. Information about the parentage of cultivars and hybrids is entered the computer memory which can be retrieved any time. The list of parents that are common in the pedigree of various cultivars and hybrids can be sorted out easily. It helps in the pedigree analysis which in turn can be used in planning plant breeding programmes especially in the selection of parents for use in hybridization programmes. Pedigree analysis is done through the study of genomics using molecular markers and protein structures.

4. Crop Modelling

Bioinformatics plays an important role in modelling of crop plants. First the plant model is conceptualized using various plant traits and then efforts are made to develop such model by using appropriate breeding procedures. First of all donor sources for targeted traits are identified from the available germplasm. These traits will be transferred in one genotype preferably in a popular variety. Such computer based studies help in developing plant ideotype suitable for machine picking and used in multiple cropping system.

5. Biometrical Analysis

In plant breeding and genetics, various types of biometrical analyses such as correlation, path coefficient, diallel, partial diallel, triallel, quadriallel, generation means, line x tester, triple test cross, stability parameters, D^2 statistics, path analysis, genetic advances, *etc.* are carried out. Computer aided programmes known as software are very much useful in carrying out such biometrical analyses.

6. Forecasting Models

Computer aided programmes have wide applications in developing various types of forecasting models especially useful for predicting crop production and productivity and in forecasting incidence of insects and diseases in crop plants. Weather parameters are used in making such predictions. Computer aided remote sensing techniques are used for such predictions.

7. Other Applications

Besides agricultural applications, bioinformatics have several other useful applications such as in medical science, forensic science, pharmaceuticals and biotech industry. In medical science computer aided studies are useful in detection of genetic diseases at an early stage of life. It can help in cure of genetic diseases in some cases, drug design and drug discovery. In forensic science, bioinformatics is useful in settling disputed cases of children and detecting criminal cases. In pharmaceutical industry, computer aided programmes help in detecting various metabolic pathways involved in the production of a medicine. Bioinformatics tools helps in sequence alignment, gene finding, genome assembly, drug design, drug discovery, protein structure alignment, protein structure prediction, prediction of gene expression, protein-protein interactions, genome-wide association studies, and the modelling of evolution.

ADVANTAGE OF BIOINFORMATICS

There are several advantages of bioinformatics. Main advantages of bioinformatics are briefly presented as follows:

1. Bioinformatics is very useful in the areas of structural genomics, functional genomics, and nutritional genomics.
2. Bioinformatics is used for transcriptome analysis where mRNA expression levels can be determined.
3. Analysis of sequencing data.
4. In rapid and accurate statistical analysis.
5. In Drug designing and QSAR modelling including similarity searching, clustering and virtual screening
6. Bioinformatics tools are very effective in prediction, analysis and interpretation of clinical and preclinical findings.
7. This information is useful in planning various breeding and genetic programmes.
8. It helps in finding evolutionary relationship between two species and in drawing phylogenetic trees.
9. Is a rapid method of gene mapping and sequencing. Earlier methods of gene mapping were time consuming and pains taking. Bioinformatics has made this task very simple. Now gene hunting has become faster, cheaper and systematic.
10. The computer based information has very high level of accuracy and is highly reliable.
11. Reconstruct genes from Expressed Sequence Tags (EST).
12. Computer aided programmes are useful in designing primers for PCR.

PRIMARY AND SECONDARY DATABASE

Due to the large volume of data that has been generated, its organization and storage become necessary. Therefore, databases were created, which constitute a large number of biological information stored and processed to allow the access to the scientific community. One of the first databases to emerge was Gene Bank, which is a collection of all available protein and DNA sequences. It is maintained by the National Institutes of Health (NIH) and the National Centre for Biotechnology Information (NCBI). NCBI was established in Maryland (US) in 1988.

Biological databases can be classified as primary, secondary, and composite databases. Primary databases contain information for sequence or structure only. Examples of primary biological databases include: Swiss-Prot and PIR for protein sequences, Gene Bank and DDBJ for genome sequences, Protein Databank for protein structures.

Secondary databases contain information derived from primary databases. Secondary databases store information such as conserved sequences, active site residues, and signature sequences. Protein Databank data is stored in secondary databases. Examples include: SCOP at Cambridge University, CATH at the University College of London, PROSITE of the Swiss Institute of Bioinformatics, e-MOTIF at Stanford. Gene Bank at the National Centre for Biotechnology Information (NCBI), DNA Database of Japan (DDBJ), and European Molecular Biology Laboratory (EMBL) stand out as the main databases of nucleotide sequences and proteins (Pevsner, 2015). As examples of secondary databases, we can point Protein Information Resource (PIR), UniProtKB/Swiss-Prot, Protein Data Bank (PDB), Structural Classification of Proteins2 (SCOP), and Prosite.

Composite databases contain a variety of primary databases, which eliminates the need to search each one separately. Each composite database has different search algorithms and data structures. The NCBI hosts these databases, where links to the Online Mendelian Inheritance in Man (OMIM) is found.

SEQUENCE ALIGNMENT AND ITS APPLICATION

Sequence alignment is widely used in molecular biology to find similar DNA or protein sequences using bioinformatics tools. These algorithms generally fall into two categories: global, which align the entire sequence, and local, which only look for highly similar sub-sequences. This Demonstration uses the Needleman–Wunsch (global) and Smith–Waterman (local) algorithms to align random English words. Gaps are shaded yellow, mismatches orange, and matches red (with a lighter shade for those matches not appearing in the final alignment). Sequence alignment is a way of arranging the sequences of DNA, RNA or protein to identify regions of similarity that may be a consequence of functional, structural or evolutionary relationships between the sequences. The procedure of comparing two (pair-wise alignment) or more multiple sequences is to search for a series of individual characters or patterns that are in the same order in the sequences.

Multiple sequence alignment (MSA) methods refer to a series of algorithmic solution for the alignment of evolutionarily related sequences, while taking into account evolutionary events such as mutations, insertions, deletions and rearrangements under certain conditions. These methods can be applied to DNA, RNA or protein sequences. There are two types of alignment: local and global. Global Alignment: Closely related sequences which are of same length are very much appropriate for global alignment. Here, the alignment is carried out from beginning till end of the sequence to find out the best possible alignment. The Needleman-Wunsch algorithm (A formula or set of steps to solve a problem) was developed by Saul B. Needleman and Christian D. Wunsch in 1970, which is a dynamic programming algorithm for sequence alignment. The dynamic programming solves the original problem by dividing the problem into smaller independent sub problems. These techniques are used in many different aspects of computer science. The algorithm explains global sequence alignment for aligning nucleotide or protein sequences. Global alignment is attempting to match as much of the sequence as possible. Local Alignment: Sequences which are suspected to have similarity or even dissimilar sequences can be compared with local alignment method. It finds local regions with high level of similarity. Local alignment is to try to find the regions with highest density of matches. The tool for local alignment is based on Smith-Waterman.

APPLICATIONS OF SEQUENCE ALIGNMENT

Multiple sequence alignment can give almost perfect protein or RNA secondary structure. Sometimes it helps even with the 3D structure.

1. Multiple sequence alignment can help to decide that your protein is a member of a known protein family or not.
2. By looking at conserved regions or sites, one can identify which region is responsible for a functional site.
3. To locate DNA regulatory elements such as binding sites
4. Phylogenetic Analysis using related sequences a tree can be constructed using sequences that you have used in the Multiple Sequence Alignment

INTRODUCTION TO BLAST AND FASTA

Among the most important algorithms used to search sequence databases at present (2003) are a family of algorithms based on BLAST, the "Basic Local Alignment Search Tool." BLAST performs particularly well with protein-coding sequences. A second, slightly older, algorithm FASTA may perform better with non-coding DNA sequences. Searching a large sequence database is a difficult problem because there are many possible ways in which the query sequence might align with the database. To speed up this process BLAST looks for small regions of perfect match between the query and target sequences, and then examines the sequence that adjoins these regions to see if there is a longer stretch that matches perfectly.

The BLAST software package is free to use (Open Source) and be installed on any local system - it's originally written for UNIX type Operating Systems. The package contains both programs for performing the actual sequence searches against pre-existing databases, *e.g.*, "blastn" for DNA databases and "blastp" for protein databases, "blast x" to compare translated nucleotide to protein, "tblastn" for comparing protein sequence to translated nucleotide.

FASTA is a DNA and protein sequence alignment software package first described by David J. Lipman and William R. Pearson in 1985. Its legacy is the FASTA format which is now ubiquitous in bioinformatics. FASTA is alignment programme for protein: protein, DNA: DNA, protein: translated DNA (with frameshifts) or vice versa. It is fast homology search tool, similar to blast but it speedup sequence comparison. In FASTA header line begins with > symbol and – will be found in gap location of particular sequence. FASTA and BLAST both tools are successfully being applied in gene mapping, phylogenetic analysis, functional analysis, metabolic pathways analysis, identification of mutation, allele mining, comparing DNA and protein sequences.

SUMMARY

The computer aided study of biology particularly genetics and molecular biology is called bioinformatics. Now the science of bioinformatics is gaining increasing importance in life science especially in the field of molecular biology and plant genetic resources.

Bioinformatics has several important applications in agriculture such as varietal identification, studies on omics, pedigree analysis, crop modelling, biometrical analysis, forecasting models, *etc.* these are briefly discussed.

Bioinformatics has several practical applications in genetics and plant breeding as discussed above. There are several advantages of bioinformatics such as systematic approach, study of evolution, rapid method, identification of similar genes, high accuracy, designing of primers and handling of huge data.

Biological databases can be classified as primary, secondary, and composite databases. All these have been explained with example. Sequence alignment is widely used in molecular biology to find similar DNA or protein sequences using bioinformatics tools. These algorithms generally fall into two categories: global, which align the entire sequence, and local, which only look for highly similar sub-sequences. A brief description of BLAST and FASTA is presented.

Computer aided programs have helped in better understanding of various processes of life science. However, there are some limitations of bioinformatics such as high initial cost, training, regular power supply, anti-virus vaccine and maintenance. These are briefly discussed.

GLOSSARY

Bioinformatics: The computer aided study of biology particularly genetics and molecular biology.

BLAST: Basic Local Alignment Search Tool. It is a DNA and protein sequence alignment software.

FASTA: Fast Adaptive Shrinkage Threshold Algorithm. It is a DNA and protein sequence alignment software.

Genomics: Study of structure and function of all genes found in a living organism.

Proteomics: Study of structure and function of all proteins found in a living organism.

Metabolomics: Study all metabolic pathways in a living organism.

QUESTIONS

1. **Define bioinformatics and describe its applications in genetics and plant breeding.**
2. **Discuss role of computers in (i) pedigree analysis, and (ii) updating of information.**
3. **Describe briefly role of computers in (i) biometrical analysis, and (ii) report writing.**
4. **Explain briefly the role of computers in handling PGR data in plant breeding.**
5. **Explain briefly the role of computers in (i) plant modeling and (ii) diagrammatic representation.**
6. **Describe briefly the role of computers in (i) maintaining varietal information system, and (ii) storage and retrieval of data.**
7. **Describe advantages and limitations of bioinformatics.**
8. **Give a brief account of primary and secondary data base with examples.**
9. **Define the following terms:**

   (i) BLAST (ii) FASTA

   (iii) Genomics (iv) Proteomics

*****

Chapter 21

Introductory Nanotechnology

INTRODUCTION

The term nanotechnology was defined in 1974 by a professor from Tokyo Science University, Norio Taniguchi. By definition nanotechnology deals with units that are smaller than 100 nanometers. Nanotechnology has been defined in various ways. Some definitions of nanotechnology are as follows:

1. Nanotechnology, in short called nanotech, is the branch of engineering that deals with things smaller than 100 nanometers (especially with the manipulation of individual molecules).
2. Nanotechnology, sometimes shortened to nanotech, refers to a field of applied science whose theme is the control of matter on an atomic and molecular scale. Generally nanotechnology deals with structures 100 nanometers or smaller, and involves developing materials or devices within that size.
3. It is the science and technology of creating nanoparticles and of manufacturing machines which have sizes within the range of 0.1 to 100 nanometer.

MAIN FEATURES

Nanotechnology deals with the creation and use of objects at the nano-scale, up to 100 nanometers in size. One nanometer is one-millionth of a millimeter or one billionth of a meter or one thousand-millionth of a meter. The main points related to nanotechnology are listed below:

1. Nanotechnology creates and uses structures, devices, and systems that have novel properties.
2. It is the manipulation or manufacture of material at the molecular level, or one billionth of a meter.

3. It includes the ability to devise self-replicating machines, robots, and computers that are molecular sized,
4. It also includes nano-delivery systems for drugs, and quantum and molecular computing.
5. It is the science and technology of building electronic circuits and devices from single atoms and molecules.
6. Nanotechnology involves working with particles at nanometer scale [at the atomic or molecular level].
7. It makes it possible to arrange atoms and molecules in certain defined structures.
8. It deals with manipulating and exploiting the properties of matter at a molecular level.
9. It is the science of particles that are measured by a nanometer, which is one billionth of a meter.

BRIEF HISTORY

The idea of the use of nanotechnology was first given in 1959 by a well-known physicist Richard Feynman in his speech at an American Physical Society meeting at California Institute of Technology (Caltech) on December 29, 1959. His speech was completely theoretical and seemingly fantastic because at that it was not possible to manipulate single atoms or molecules as they were far too small for manipulation.

He described that the laws of physics do not limit our ability to manipulate single atoms and molecules. However, it was our lack of the appropriate methods for doing so. He correctly predicted that the time would come in which atomically precise manipulation of matter would inevitably arrive.

APPLICATIONS

Nanotechnology has a number of potential applications in diversity of fields such as (i) medicine, (ii) filtration, (iii) energy, (iv) displays, (v) consumer goods, (vi) agriculture, and (vii) crop improvement. First applications of nanotechnology have been discussed in crop improvement and agriculture and then applications in other fields are presented in Table 21.1.

TABLE 21.1: Applications of Nanotechnology

Sl.No.	*Main Field*	*Applications*
1	Agricultural Biotechnology	(i) Gene delivery
	[Crop Improvement]	(ii) Fine control
		(iii) Multiple gene delivery
		(iv) Study of gene function

Sl.No.	*Main Field*	*Applications*
2	Agriculture	(i) Detection of disease (ii) Treatment of disease (iii) Better nutrient absorption (iv) Better use of insecticides and herbicides (v) Protection of environment (vi) Precision farming
3	Medicine	(i) Drug delivery (ii) Tissue engineering:
4	Filtration	Waste-water treatment, air purification
5	Energy	(i) Increasing the efficiency of energy production (ii) Recycling of batteries
6	Displays	Carbon nanotubes
7	Consumer goods	Foods, house hold, optics, textiles and cosmetics.

A. Application in Crop Improvement

Nanotechnology has practical applications in crop improvement through its use in agricultural biotechnology. A team of Iowa State University plant scientists and chemists [Brian Trewyn, Francois Torney, Kan Wang and Victor Lin] first used nanotechnology to penetrate rigid plant cell walls and deliver DNA and chemicals with precise control. They successfully used nanotechnology to penetrate plant cell walls and deliver a gene and a chemical that triggers its expression. This application of nanotechnology to plant biology and agricultural biotechnology has resulted in a significant break-through in delivering specific gene into plant cells. The main applications of nanotechnology in agricultural biotechnology include (i) gene delivery, (ii) controlled gene delivery, (iii) multiple gene delivery, and (iv) study of gene function. These are briefly discussed below:

1. **Gene Delivery**: It permits successful delivery of a gene into a plant cell. Till recently, we were at nature's mercy when we delivered a gene into a cell," Lin said. "There has been no precise control as to whether the cells will actually incorporate the gene and express the consequent protein. With this technology, we may be able to control the whole sequence in the future." And once the gene is inside the plant cell wall, it opens up "whole new possibilities. In the future, scientists could use the new technology to deliver imaging agents or chemicals inside cell walls. This would provide plant biologists with a window into intracellular events.
2. **Controlled Technique**: We can bring in a gene and induce it in a controlled manner at the same time and at the same location. That has never been done before" [Wang}.
3. **Multiple Gene Delivery:** The most tremendous advantage is that several genes can be delivered into a plant cell at the same time and released them whenever required. With the mesoporous Nano-particles, it is possible to deliver two biogenic species at the same time.

4. **Study of Gene Function:** The controlled release of gene in to plant cell will improve the ability to study gene function in plants. In a separate process, chemicals are used to activate the gene's function. The process is imprecise and the chemicals could be toxic to the plant.

In plant, transformation mostly occurs with the use of a gene gun. In order to use the gene gun to introduce the nanoparticles to walled plant cells, the chemists made another useful modification on the particle surface. They synthesized even smaller gold particles to cap the nanoparticles. These "golden gates" not only prevented chemical leakage, but also added weight to the nanoparticles, enabling their delivery into the plant cell with the standard gene gun.

Achievements

The biologists have successfully used the nanotechnology to introduce DNA and chemicals to *Arabidopsis*, tobacco and corn plants.

The team also found a chemical that made the nanoparticle look yummy to the plant cells so they would swallow the particles. The nanoparticles were swallowed by the plant protoplasts [naked cells].

B. Applications in Agriculture

There are new challenges in agricultural sector. Such challenges include a growing demand for healthy and safe food, an increasing risk of disease, and threats to agricultural and fishery production from changing weather patterns. Nanotechnology has the potential to revolutionize the agricultural and food industry with new tools. Important applications of nanotechnology in the field of agriculture include (i) treatment of diseases, (ii) detection of diseases, (iii) better nutrient absorption, (iv)better use of insecticides and herbicides, and (v) protection of environment. All these have been discussed below:

1. **Treatment of Diseases:** It provides molecular treatment of diseases.
2. **Detection of Diseases:** It helps in rapid detection of diseases. Smart sensors and smart delivery systems will help the agricultural industry combat viruses and other crop pathogens.
3. **Better Nutrient Absorption:** It enhances the ability of plants to absorb nutrients *etc.*
4. **Better Use of Insecticides and Herbicides:** In the near future nano structured catalysts will be available which will increase the efficiency of pesticides and herbicides, allowing lower doses to be used.
5. **Protection of Environment:** Nano-technology will also protect the environment indirectly through the use of alternative (renewable) energy supplies, and filters or catalysts to reduce pollution and clean-up existing pollutants.

   An agricultural methodology widely used in the USA, Europe and Japan, which efficiently utilizes modern technology for crop management, is called Controlled Environment Agriculture (CEA). CEA is an advanced and intensive form of hydroponically-based agriculture. Plants are grown within a controlled environment so that horticultural practices can be optimized. The computerized system monitors and regulates localized environments such as fields of crops. CEA technology, as it exists today, provides an excellent platform for the introduction of nanotechnology to agriculture. In CEA, nanotechnological devices can tremendously improve the grower's ability to determine the best time of harvest for the crop, the vitality of the crop, and food security issues, such as microbial or chemical contamination.

6. **Precision Farming:** There are two main applications of nanotechnology in the precision farming, *viz.* (i) reduction in the cost of production, and (ii) reduction in pollution. These are discussed below:
   (i) Reduction in Cost of Production: The major goal of precision farming is to maximize output (*i.e.* crop yields) and minimize input (*i.e.* fertilizers, pesticides, herbicides, *etc.*) by monitoring environmental variables and applying targeted action. Precision farming makes use of computers, global satellite positioning systems, and remote sensing devices to measure highly localized environmental conditions. Thus it determines whether crops are growing at maximum efficiency or precisely identifies the nature and location of problems. By using centralized data to determine soil conditions and plant development, seeding, fertilizer, chemical and water use can be fine-tuned to lower production costs and potentially increase production- all benefiting the farmer.
   (ii) Reduction in Pollution: Precision farming can also help to reduce agricultural waste and thus keep environmental pollution to a minimum. Although not fully implemented yet, tiny sensors and monitoring systems enabled by nanotechnology will have a large impact on future precision farming methodologies. One of the major roles for nanotechnology-enabled devices will be the increased use of autonomous sensors linked into a GPS system for real-time monitoring. These Nano sensors could be distributed throughout the field where they can monitor soil conditions and crop growth. Wireless sensors are already being used in certain parts of the USA and Australia. For example, one of the Californian vineyards, Pick-berry, in Sonoma County has installed wifi systems with the help of the IT Company, Accenture. The initial cost of setting up such a system is justified by the fact that it enables the best grapes to be grown which in turn produce finer wines, which command a premium price. The use of such wireless networks is of course not restricted to vineyards only. This technique can also be used in other crops.

C. Other Applications

Besides above, nanotechnology has wide practical applications in a diversity of fields such as (i) medicine, (ii) filtration, (iii) energy, (iv) displays, (v) consumer goods. These are briefly presented below:

1. Medicine

(i) **Drug delivery:** The overall drug consumption and side-effects can be lowered significantly by depositing the active agent in the morbid region only and in no higher dose than needed. This highly selective approach reduces costs and human suffering. An example can be found in dendrimers and Nano-porous materials. They could hold small drug molecules and transporting them to the desired location.

(ii) **Tissue engineering:** Nanotechnology can help to reproduce or to repair damaged tissue. Tissue engineering might replace today's conventional treatments, *e.g.*, transplantation of organs or artificial implants.

2. Filtration

A strong influence of nano-chemistry on waste-water treatment, air purification and energy storage devices is to be expected. Nano-filtration is mainly used for the removal of ions or the separation of different fluids.

Magnetic nano-particles offer an effective and reliable method to remove heavy metal contaminants from waste water by making use of magnetic separation techniques. Using nano-scale particles increases the efficiency to absorb the contaminants and is comparatively inexpensive compared to traditional precipitation and filtration methods.

3. Energy

(i) **Increasing the efficiency of energy production:** Nanotechnology could improve combustion by designing specific catalysts with maximized surface area. Scientists have recently developed tetrad-shaped Nano-particles that, when applied to a surface, instantly transform it into a solar collector.

(ii) **Recycling of batteries:** The use of batteries with higher energy content or the use of rechargeable batteries or super-capacitors with higher rate of recharging using nano-materials could be helpful for the battery disposal problem.

4. Displays

The production of displays with low energy consumption could be accomplished using carbon nanotubes (CNT). Carbon nanotubes can be electrically conductive and due to their small diameter of several nanometers, they can be used as field emitters with extremely high efficiency.

5. Consumer Goods

In case of consumer goods, nanotechnology has wide applications in foods, house hold, optics, textiles and cosmetics as discussed below:

(i **Foods:** Nanotechnology can be applied in the production, processing, safety and packaging of food. A Nano-composite coating process could improve food packaging by placing anti-microbial agents directly on the surface of the coated film. Nano-composites could increase or decrease gas permeability of different fillers as is needed for different products.

They can also improve the mechanical and heat-resistance properties and lower the oxygen transmission rate. Research is being performed to apply nanotechnology to the detection of chemical and biological substances for sensing biochemical changes in foods.

(ii) Household: The most prominent application of nanotechnology in the household is self-cleaning or easy-to-clean surfaces on ceramics or glasses. Nano-ceramic particles have improved the smoothness and heat resistance of common household equipment.

(iii) Optics: The first sunglasses using protective and antireflective ultrathin polymer coatings are on the market. For optics, nanotechnology also offers scratch resistant surface coatings based on Nano-composites. Nano-optics could allow for an increase in precision of pupil repair and other types of laser eye surgery.

(iv) Textiles: The use of engineered nano-fibers already makes clothes water- and stain-repellent or wrinkle-free. Textiles with a Nano-technological finish can be washed less frequently and at lower temperatures. Nanotechnology has been used to integrate tiny carbon particles membrane and guarantee full-surface protection from electrostatic charges for the wearer. A military application could be in camouflage where Nano-cameras mixed with nano-displays could create an invisibility coat, acting like the skin of a Chameleon.

(v) Cosmetics: One field of application is in sunscreens. The traditional chemical UV protection approach suffers from its poor long-term stability. A sunscreen based on mineral nanoparticles such as titanium dioxide offer several advantages. Titanium oxide nanoparticles have a comparable UV-protection property as the bulk material, but lose the cosmetically undesirable whitening as the particle size is decreased.

IMPLICATIONS OF NANOTECHNOLOGY

Nanotechnology has a number of potential useful applications, but the side effect of this advance technology can not be ignored. The effects of Nano-materials on human health and environment are briefly presented below:

1. Effects on Human Health

There are scientific evidences which demonstrate that some Nano-materials are toxic to humans or the environment. The smaller a particle, the greater is its surface area to volume ratio and the higher is its chemical reactivity and biological activity. nano-materials toxicity may result in oxidative stress, inflammation, and consequent damage to proteins, membranes and DNA.

The extremely small size of nano-materials also means that they are much more readily taken up by the human body than larger sized particles. Size is therefore a key factor in determining the potential toxicity of a particle. However, it is not the only important factor. Other properties of nano-materials that influence toxicity include: chemical composition, shape, surface structure, surface charge, aggregation and solubility, and the presence or absence of functional groups of other chemicals.

The health risks of nano-materials are of particular concern for workers who may face occupational exposure to nano-materials at higher levels, and on a more routine basis, than the general public.

2. Effects on Environment

The disposal of Nano-particles will pollute the environment and affect the ecosystem. Hence, the authorities should take appropriate action to minimize the side effects of Nano-materials. There should be awareness in the public about the side effects of nanotechnology.

SUMMARY

The term nanotechnology was defined in 1974 by a professor from Tokyo Science University, Norio Taniguchi. By definition nanotechnology deals with units that are smaller than 100 nanometers. Nanotechnology has been defined in various ways. Some definitions of nanotechnology are as follows:

It is the branch of engineering that deals with things smaller than 100 nanometers (especially with the manipulation of individual molecules).

Nanotechnology deals with the creation and use of objects at the Nano scale, up to 100 nanometers in size. One nanometer is one-millionth of a millimeter or one billionth of a meter or one thousand-millionth of a meter. The main points related to nanotechnology have been discussed.

Nanotechnology has a number of potential applications in diversity of fields such as (i) medicine, (ii) filtration, (iii) energy, (iv) displays, (v) consumer goods,(vi) agriculture, and (vii) crop improvement. All these have been discussed.

Important applications of nanotechnology in the field of agriculture include (i) treatment of diseases, (ii) detection of diseases, (iii) better nutrient absorption, (iv) better insecticides and herbicides use, and (v) protection of environment. All these have been discussed.

There are two main applications of nanotechnology in the precision farming, *viz.* (i) reduction in the cost of production, and (ii) reduction in pollution. These have been discussed.

The main applications of nanotechnology in agricultural biotechnology include (i) gene delivery, (ii) controlled gene delivery, (iii) multiple gene delivery, and (iv) study of gene function. All these have been discussed.

Applications in the field of (i) medicine, (ii) filtration, (iii) energy, (iv) displays, (v) consumer goods have been briefly presented.

Nanotechnology has a number of potential useful applications, but the side effects of this advance technology can not be ignored. The effects of Nano-materials on human health and environment have been discussed.

GLOSSARY

Gene Delivery: Successful transfer of a gene into a plant cell.

Nanotechnology: The branch of engineering that deals with things smaller than 100 nanometers.

Nano Particles: Structures of 100 nanometers or smaller.

Precision Farming: Accurate farming.

QUESTIONS

1. **Define nanotechnology and describe its main features.**
2. **Describe briefly various applications of nanotechnology.**
3. **Explain briefly applications of nanotechnology in agriculture.**
4. **What are the important applications of nanotechnology in crop improvement?**
5. **Discuss briefly general applications of nanotechnology.**
6. **Explain applications of nano technology in the field of:**
   (*a*) Medicine
   (*b*) Agriculture
   (*c*) Crop Improvement
   (*d*) Textiles
7. **Describe briefly side effects of nanotechnology.**
8. **Explain nanotechnology in relation to precision farming.**
9. **Write short notes on the following:**
   (*a*) Nano-technology
   (*b*) Nano-particles
   (*c*) Nanometer
   (*d*) Nano-engineering

*****

Chapter 22

Biosafety Regulations

INTRODUCTION

The Cartagena Protocol on Bio-safety is the first international agreement to regulate the trans-boundary movements of genetically engineered (GE) organisms. The Bio-safety Protocol is a subsidiary agreement to the UN Convention on Biological Diversity (CBD), which was signed by over 150 governments at the Rio Earth Summit in 1992. The Bio-safety Protocol seeks to protect biological diversity from the potential risks posed by living modified organisms [LMOs] resulting from modern biotechnology.

Bio-safety refers to the safe application of modern biotechnology with regard to human health, animal health and environment. In other words, bio-safety is a term used to describe efforts to reduce and eliminate potential risk resulting from biotechnology and its products.

IMPORTANT POINTS

Biosafety is an issue of global significance. Important points related to biosafety measures are briefly presented as follows:

1. **Application:** Bio-safety is applicable to the food industry, public health, agriculture and environment. In other words, it is related to agriculture, medicines, ecology, chemistry, *etc.* It is also applied to crop production, conservation, marketing and trade. It is required in World Trade Organization, FAO and WHO on traded food substances.
2. **Organizations:** Many developed countries have established their domestic bio-safety systems. At International level, Cartegina Protocol has been established to look into biosafety aspects.
3. **Advantages:** Bio-safety measures reduce the risk of health hazards to laboratory workers, persons outside the laboratory and the environment from the infectious agents.

4. **Levels of Bio-safety:** There are four levels of biosafety, *viz.* level 1, level 2, level 3 and level 4. See later.
5. **Biosecurity:** It refers to prevention of hazards to human health, animal health and environment from artificial pathogens (bacteria, fungi and viruses) used in biological warfare and robots.

BIOSAFETY LEVELS

Biosafety level refers to stringency of precautions to be adopted by Research Centers for laboratory work. In other words, it refers to safety from exposure to infectious agents.

As stated above, there are four levels of biosafety, *viz.* level 1, level 2, level 3 and level 4. A brief description of each level of biosafety is presented as follows:

1. Bio-safety Level 1

It is suitable for work involving agents of minimal potential hazards to laboratory workers and environment. Other points are given below:

(i) Laboratory is not isolated from the general building.
(ii) The work is done generally on open bench top using standard microbiological techniques.
(iii) Special containment equipment generally is not used.
(iv) Well-trained personals work in the laboratory.
(v) Workers are supervised by a well-trained Scientist in microbiology in a related field.

2. Bio-safety Level 2

It is suitable for work involving agents of moderate potential hazards to laboratory personas and environment. Other points are given below:

(i) Laboratory is not isolated from the general building.
(ii) Special containment equipment is sometimes required.
(iii) Well-trained personals work in the laboratory for handling pathogenic agents.
(iv) Workers are supervised by a competent Scientist.
(v) Extreme precautions are with contained sharp items.

3. Bio-safety Level 3

It is suitable for work involving agents of high potential hazards to laboratory workers and the environment. Such agents may cause serious or lethal disease to the workers by inhalation. Other points are given below:

(i) Laboratory is isolated from the general building.
(ii) The laboratory has special engineering design.
(iii) Workers are supervised by a competent Scientist.

(iv) Workers wear appropriate protective clothing and equipment.
(v) Access to laboratory is restricted when work is in progress.
(vi) Well-trained personals work in the laboratory.

4. Bio-safety Level 4

It is required for work involving dangerous agents with very high risk aerosol transmitted infections and life threatening disease. Other points are given below:

(i) Laboratory is completely isolated from other areas of general building.
(ii) Laboratory workers have special training in handling extremely hazardous infectious agents.
(iii) Workers are supervised by a competent and experienced Scientist.
(iv) All procedures are conducted in physical containment equipment or devices.
(v) Access to laboratory is strictly restricted by the laboratory Director.
(vi) Laboratory has special engineering design to prevent micro-organisms from being disseminated into the environment.

BIOSAFETY PROTOCOL

The Cartagena Protocol on Bio-safety, also known as the Bio-safety Protocol, is a supplement of the convention on biological diversity. The Bio-safety Working Group met six times between 1996 and 1999. The extraordinary meeting of the conference of parties [COP], was originally scheduled to adopt the Protocol in February 1999 in Cartagena city of Colombia. However, the meeting broke down because of disagreements regarding appropriate consideration of economic interests. The extraordinary meeting was temporarily suspended and only resumed after three more negotiation rounds achieved compromise. The Protocol was finally adopted by conference of parties [COP] in Montreal, Canada on 29 January 2000 and entered force on 11 September 2003.

The need of such a strong Bio-safety Protocol is illustrated by the genetic contamination of maize in Mexico. Genetically engineered cultivars of maize in Mexico resulted in contamination of non-transgenic varieties, land races and wild relatives of this major food crop. This was the first case of genetic pollution in a centre of origin and diversity of a major food crop. This invited the attention of crop scientists for adoption of strong bio-safety protocol for handling of LMOs.

Member Countries

By June 2001, the Protocol had received 103 signatures. The membership is open to all countries of the World. The membership has been signed by 162 countries till November 2011. India signed the Cartagena Protocol in January 2001. However, countries like USA, Argentina and Canada that produce about 90 percent of genetically engineered [GE] crops in the world have not ratified the Protocol, and are actively working to undermine it.

Entry into Force

The required number of 50 instruments of ratification/accession/approval/acceptance by countries was reached in May 2003. As a result, the Protocol entered into force on 11 September 2003.

Governing Body

The governing body of the Protocol is the Conference of the Parties to the Convention serving as the meeting of the Parties to the Protocol (COP-MOP). The main function of this body is to review the implementation of the Protocol and make decisions necessary to promote its effective operation. As of date, four meetings of the COP-MOP have been convened. The governing body includes Chairman and members from member countries. The Chairman is of the rank of Ambassador or High Commissioner of the member country. The term of the governing body is usually five years. The secretariat of the protocol is the same as for convention on biological diversity [CBD]. It is located at Montreal in Canada. Funds for the protocol are voluntarily contributed by contracting parties, developed country parties and other countries and sources. The main functions of governing body are:

1. To review the implementation of the Protocol and
2. To make decisions necessary to promote its effective operation.

FUNCTIONS OF BIOSAFETY PROTOCOL

Bio-safety protocol refers to various measures that are taken for the safe application of modern biotechnology. These are regulatory measures and policies that are designed to safeguard the public interest. The bio-safety protocol became effective from January 2000. The main functions of bio-safety protocol are as follows:

1. **Promotion of Bio-safety:** The Protocol promotes bio-safety measures by establishing rules and procedures for the safe transfer, handling, and use of LMOs, developed through modern biotechnology.
2. **Conservation of Biodiversity:** The Protocol applies to the trans-boundary movement [transit, handling and use] of all living modified organisms that may have adverse effects on the conservation and sustainable use of biological diversity. It ensures that LMOs should not have adverse effects on human health, animal health and biological diversity.
3. **Regulation of World Trade:** Protocol is designed to regulate the international trade of genetically engineered organisms that may have adverse effects on the conservation and sustainable use of biological diversity, including risk to human health.

OBJECTIVE AND ELEMENTS OF PROTOCOL

The objective of the Protocol is to ensure an adequate level of protection in the field of the safe transfer, handling and use of 'living modified organisms resulting from modern biotechnology' that may have adverse effects on the conservation and sustainable use of biological diversity, including risks to human health, specifically from trans-boundary movement of LMOs. The main elements of Cartagena biosafety protocol are briefly presented here as under.

1. **Clearing House:** The Protocol establishes an Internet-based "Biosafety Clearing-House" to help countries exchange scientific, technical, environmental, and legal information about living modified organisms (LMOs).
2. **Advance Informed Agreement Procedure:** It creates an advance informed agreement (AIA) procedure that in effect requires exporters to seek consent from an importing country before the first shipment of an LMO meant to be introduced into the environment (such as seeds for planting, fish for release, or microorganisms for bioremediation).

   The AIA procedure does not apply to LMO commodities that are intended for food, feed, or processing (*e.g.*, maize, soy or cottonseed), to LMOs in transit, or to LMOs destined for contained use (*e.g.*, organisms intended only for scientific research within a laboratory).
3. **Items not addressed**: The Protocol does not address food safety issues. It does not pertain to non-living products derived from genetically engineered plants or animals, such as milled maize or other processed food products.
4. **Documentation:** It requires shipments of LMO commodities, such as maize or soybeans that are intended for direct use as food, feed, or for processing, to be accompanied by documentation stating that such shipments "may contain" living modified organisms and are "not intended for intentional introduction into the environment". It also sets out information to be included on documentation accompanying LMOs destined for contained use, including any handling requirements and contact points for further information and name and address of the importer and exporter. It should also include a declaration that the movement is in conformity with the Protocol and, as appropriate.
5. **Existing Rights Remain Unchanged:** Parties must implement rights and obligations under the Protocol consistent with their existing international rights [such as WTO agreements] and obligations, including with respect to non-Parties to the Protocol.
6. **Capacity Building:** The Protocol calls on Parties to cooperate with developing countries in building their capacity for managing modern biotechnology.
7. **Trade with non-Parties:** The Protocol states that the "trans-boundary movement of living modified organisms between Parties and non-Parties shall be consistent with the objective of this Protocol.

RISK ASSESSMENT

There are two broad categories of risk from genetically modified organisms and their products, *viz.*, (i) risk to human and animal health, and (ii) risk to the environment. These aspects are briefly discussed as follows:

1. Risk to Human and Animal Health

The risk from genetically modified organisms and their products to human health include toxicity, allergenicity, antibiotic resistance and nutritional uptake. These are discussed as follows:

(i) **Toxicity:** Sometimes, the transgene leads to production of toxic substance by altering the metabolic pathway. Consumption of such transgenic products may cause toxic effect on human health. Hence, every transgenic product has to be evaluated for toxicity to human and animal health.

(ii) **Allergenicity:** Sometimes, transgenic products have allergic effects on human health. For example, transgenic soybean containing methionine gene from Brazil nut has allergic effects in sensitive persons. Hence, such soybean has not been approved for sale. Feeding of *Bt.* cotton seed to animals has not been reported to have any adverse effect. The *Bt.* cake also does not have any adverse effect on the digestion of animals. Moreover, no allergic or toxic effect of Bt. cotton seed and meal has been reported so far. The oil extracted from the seed of *Bt.* cotton has not been found to have any adverse effect on the human health.

(iii) **Antibiotic Resistance:** Antibiotic such as Kanamycin is used for selection of transgenic cells. It is believed that transgenic products when consumed will lead to resistance to such antibiotic. However, there is no such report so far.

(iv) **Nutrient Uptake:** There are possibilities that transgene product may inhibit absorption of some nutrients in human digestive system, resulting in adverse effects on human health. However, such evidences are not available.

2. Risk to Environment

The risk from genetically modified organisms and their products to environment include development of super weeds, loss of biodiversity, contamination of non-transgenic cultivars, effect on non-target organisms, development of resistance, effect on soil ecology and evolution of new viruses. These aspects are briefly discussed as follows:

(i) **Development of Super Weeds:** Natural out crossing of transgenic plants with wild species and relatives of a crop may lead to development of super weeds. Such weeds may have herbicide resistance or insect resistance and will be more invasive than normal weeds. The out crossing is easy with other varieties or cross compatible species.

(ii) **Loss of Biodiversity:** Some farmers still maintain landraces and some old varieties which are valuable genetic resources. The natural ecosystem is still rich in plant biodiversity which needs to be conserved. The wide spread use of transgenic cultivars of different crops will lead to substantial loss of crop biodiversity.

(iii) Contamination: The cultivation of transgenic varieties will lead to contamination of landraces, non-transgenic cultivars and closely related wild species through natural out crossing. In Mexico, great diversity of both wild and cultivated corn is found. It has been reported that wild corn varieties in some remote areas of Mexico has been contaminated by Bt. transgenic cultivars.

The possibilities of cross pollination of Bt. cotton with other species are nil to negligible because the *Bt.* gene has been inserted in upland cotton (2n = 52) which cannot outcross with cultivated or wild diploid species (2n = 26). It can also not outcross with tetraploid wild species such as *G. tomentosum* which are found either in uncultivated areas or extremely isolated species garden maintained at different research institutes. The upland cotton in which *Bt.* gene has been inserted does not have cross compatibility with other genera of the family *Malvaceae.*

(iv) Effect on Non-target Insects: The genetically modified genes may sometimes have adverse effects on non-target insects/organisms. The adverse effects of *Bt.* insecticidal protein have been reported. No adverse effect of *Bt.* cotton has so far been reported on parasites, predators and other non-target beneficial insects such as honey bees, silk worm and Lac worm. However some adverse effects of *Bt.* gene have been reported on Monarch butterfly, Lice wings (an insect predator) and tobacco horn worm.

(v) Development of Resistance: There is a fear that over a period of time the target insect may develop resistance to *Bt.* insecticidal gene. However, in cotton there is no report on development of resistance to target insect so far. The first report of development of resistance to *Bt.* toxin is that of diamond back moth, an important pest of *Brassica* crops the world over.

(vi) Effects on Soil Ecology: It is considered that *Bt.* crops will liberate some additional chemicals from their roots and have adverse effects on soil ecology. *Bt.* [cry1 Ac] gene has been used in many crops, but no adverse effects of this gene on soil ecology has been reported so far.

(vii) Evolution of New Viruses: There is fear that development of virus resistant cultivars may sometimes lead to development of new stronger variants of virus that can affect transgenic plants.

REGULATORY MECHANISMS

India has a well-defined regulatory mechanism for development and evaluation of genetically modified organisms [GMOs] and their products. The Department of Biotechnology [DBT] and the Ministry of Environment and Forests [MoEF] are the two apex regulatory bodies that look into bio-safety aspects of GMOs. Guidelines for safety have been issued by the Department of Biotechnology [DBT] in 1990 covering research in biotechnology, field trials and commercial applications. Presently, there are following six competent authorities for implementation of regulations and guidelines in the country.

1. Recombinant DNA Advisory Committee [RDAC],
2. Review Committee on Genetic Manipulation [RCGM],
3. Genetic Engineering Approval Committee [GEAC]- an apex body,
4. Institutional Bio-safety Committee [IBSC] attached to every organization engaged in recombinant DNA research,
5. State Bio-safety Coordination Committees [SBCC}, and
6. District Level Committees [DLC].

A brief description of the activities of the above committees is presented below.

1. Recombinant DNA Advisory Committee [RDAC]

This committee is constituted by the Department of Biotechnology. It takes note of development in biotechnology at national and International levels. RDAC prepares recommendations from time to time that are suitable for implementation for upholding the safety regulations in research and applications of GMOs and their products. This committee prepared the Recombinant DNA Bio-safety Guidelines in 1990, which was adopted by the Government of India for conducting research and handling of GMOs in India.

2. Review Committee on Genetic Manipulation [RCGM]

The RCGM is constituted by DBT for a period of three years. It consists of chairman, co-chairman, member secretary, co-member secretary and 18-20 members from different research organizations. The RCGM works under the Department of Biotechnology and has the following functions:

(i) To bring out manuals of guidelines specifying producers for regulatory process on GMOs in research, use and applications including industry with a view to ensure environmental safety.

(ii) To review all ongoing r-DNA projects involving high risk, category and controlled field experiments.

(iii) To lay down procedure for restriction or prohibition, production, sale, import and use of GMOs both for research and applications.

(iv) To permit experiments with category II risks and above with appropriate containment.

(v) To authorize imports of GMOs/transgenes for research purposes.

(vi) To authorize field experiments in 20 acres in multi-location in one crop season with up to one acre at one site.

(vii) To generate relevant data on transgenic material in appropriate systems.

(viii) To undertake visits of sites of experimental facilities periodically, where projects with biohazard potentials are being pursued and also at a time prior to the commencement of the activity to ensure that adequate safety measures are taken as per the guidelines.

3. Genetic Engineering Approval Committee [GEAC]

It functions as a body under Ministry of Environment and Forests and is responsible for approval of activities involving large-scale use of hazardous micro-organisms and recombinant products in research and industrial production from the environment angle. Its main functions are as follows:

(i) To permit the use of GMOs and their products for commercial application.

(ii) To adopt producers for production, sale, import and use of GMOs both for research and applications under Environment Protection Authority [EPA].

(iii) To authorize large-scale production and release of GMOs and their products into the environment.

(iv) To authorize agencies or persons to have powers to take punitive action under the EPA.

4. Institutional Bio-safety Committee [IBSC]

The IBSC is the nodal point for interaction with institution for implementation of the guidelines. It is constituted before undertaking any project involving manipulation of micro-organisms, plants or animals. It consists of Head of institution, 3 or more scientists working in molecular biology, one outside expert in the discipline and one member nominated by DBT. The main activities of IBSC are as follows:

(i) To note and approve r-DNA work.

(ii) To ensure adherence of r-DNA safety guidelines of government.

(iii) To prepare emergency plan according to guidelines.

(iv) To recommend to RCGM about category II risk or above experiments and to seek approval of RCGM.

(v) To act as a nodal point for interaction with statutory bodies.

(vi) To ensure experimentation at designated locations, taking into account approved protocol.

5. State Bio-safety Coordination Committees[SBCC}

This committee is constituted in each state where research and application of GMO is carried out. It is headed by Chief Secretary of the State. He can nominate State Government representative in the activities related to field inspection of GMOs. Main functions of this committee are as follows:

(i) To ensure adherence to r-DNA safety guidelines issued by the government.

(ii) To review periodically the safety and control measures in various institutions handling GMOs.

(iii) To act as nodal agency at State level to assess the damage, if any, due to release of GMOs and to take on site control measure.

(iv) To coordinate activities related to GMOs in the state with central Ministries.

6. District Level Committees [DLC]

This is constituted at the district level and is the smallest committee. It is headed by the District Collector. Its functions are as follows:

(i) To monitor the safety regulations in the institutions.

(ii) It has powers to inspect, investigate and report to the SBCC or GEAC about compliance or non- compliance of r-DNA guidelines or violations under EPA.

(iii) To act as nodal agency at District level to assess the damage, if any, due to release of GMOs and to take on site control measures.

SUMMARY

Bio-safety refers to the safe application of modern biotechnology with regard to human health, animal health and environment. Brief history and need for Cartagena bio-safety protocol are given. By June 2001, the Protocol had received 103 signatures. The membership is open to all countries of the World. The Protocol entered into force on 11 September 2003. The membership has been signed by 162 countries till November 2011. India signed the Cartagena Protocol in January 2001. However, countries like US, Argentina and Canada that produce about 90 per cent of genetically engineered [GE] crops in the world have not ratified the Protocol, and are actively working to undermine it.

The governing body of the Protocol is the Conference of the Parties to the Convention serving as the meeting of the Parties to the Protocol (COP-MOP). The main function of this body is to review the implementation of the Protocol and make decisions necessary to promote its effective operation. The secretariat of the protocol is the same as for convention on biological diversity [CBD]. It is located at Montreal in Canada. Funds for the protocol are voluntarily contributed by contracting parties, developed country parties and other countries and sources. The objective and main elements of Cartagena bio-safety protocol are briefly presented.

There are two broad categories of risk from genetically modified organisms and their products, *viz.*, (i) risk to human and animal health, and (ii) risk to the environment. Risk to human health includes toxicity, allergenicity, antibiotic resistance and nutritional uptake. The risk to environment include development of super weeds, loss of biodiversity, contamination of non-transgenic cultivars, effect on non-target organisms, development of resistance, effect on soil ecology and evolution of new viruses.

In India, there are six competent authorities for implementation of bio-safety regulations and guidelines. These are: Recombinant DNA Advisory Committee [RDAC], Review Committee on Genetic manipulation [RCGM], Genetic Engineering Approval Committee [GEAC]- an apex body, Institutional Bio-safety Committee [IBSC] attached to every institution engaged in recombinant DNA research, State Bio-safety Coordination Committees[SBCC}, and District Level Committees [DLC]. Roles of all these committees have been discussed.

GLOSSARY AND ABBREVIATIONS

Bio-safety: The safe application of modern biotechnology with regard to human health, animal health and environment.

Bio-security: Prevention of hazards to human health, animal health and environment from artificial bacteria and viruses used in biological warfare.

AIA: Advanced Information Agreement.

CBD: Convention on Biological Diversity

COP: Conference of Parties

GMOs: Genetically Modified Organisms.

LMOs: Living Modified Organisms

MOP: Meeting of Parties.

QUESTIONS

1. **Define biosafety and describe briefly important points related to biosafety.**
2. **What is Cartagena protocol? Explain briefly its history and objectives.**
3. **Give a brief account of various biosafety levels.**
4. **Describe in brief functions of Cartagena protocol.**
5. **Write short notes on the following:**
   (i) Need for Cartagena protocol
   (ii) Member of the governing body of Cartagena protocol.
6. **What are the risk from GMOs to the human health and animal health?**
7. **Describe briefly risks from GMOs to the environment.**
8. **Discuss briefly bio-safety regulations related to GMOs in India.**
9. **Explain the role of following committees:**
   (i) Recombinant DNA Advisory Committee[RDAC],
   (ii) Review Committee on Genetic manipulation [RCGM],
   (iii) Genetic Engineering Approval Committee [GEAC].
10. **Describe briefly the functions of the following committees.**
   (i) Institutional Bio-safety Committee [IBSC].
   (ii) State Bio-safety Coordination Committee [SBCC}.
   (iii) District Level Committees [DLC].
11. **Define the following terms:**
   (i) Biosecurity (ii) Bio-safety
   (iii) Allergen-city (iv) Pathogenicity

*****

Appendices

Appendix 1
List of Indian Agricultural Universities having Plant Biotechnology Centre

A: Agricultural Universities

Sl.No.	*Name of University*	*Location*	*State*
1	Acharya NG Ranga Agricultural University	Hyderabad	Telangana
2	Assam Agricultural University	Jorhat	Assam
3	Bihar Agricultural University	Sabour	Bihar
4	Birsa Agricultural University	Ranchi	Jharkhand
5	Indira Gandhi Krishi Vishwavidyalaya	Raipur	Chhatishgarh
6	Anand Agricultural University	Anand	Gujarat
7	Junagadh Agricultural University	Junagarh	Gujarat
8	Navsari Agricultural University	Navsari	Gujarat
9	Sardar Krushinagar-Dantiwada Agricultural University	Bansakantha	Gujarat
10	Chaudhary Charan Singh Haryana Agricultural University	Hisar	Haryana
11	CSK Himachal Pradesh Krishi Vishvavidyalaya,	Palampur	Himachal Pradesh
12	Sher-e-Kashmir University of Agricultural Sciences and Technology	Jammu	Jammu and Kashmir
13	Sher-e-Kashmir University of Agricultural Sciences and Technology of Kashmir	Srinagar	Jammu and Kashmir
14	University of Agricultural Sciences	Bangalore	Karnataka
15	University of Agricultural Sciences	Dharwad	Karnataka
16	University of Agricultural Sciences	Raichur	Karnataka
17	Kerala Agricultural University	Vallanikara	Kerela
18	Jawaharlal Nehru Krishi Viswavidyalaya	Jabalpur	Madhya Pradesh
19	Rajmata Vijayaraje Scindia Krishi Viswavidyalaya	Gwalior	Madhya Pradesh
20	Dr Balasaheb Sawant Konkan Krishi Vidyapeeth	Ratnagiri	Maharashtra
21	Dr Panjabrao Deshmukh Krishi Vidyapeeth	Akola	Maharashtra
22	Mahatma Phule Krishi Vidyapeeth	Rahuri	Maharashtra
23	Marathwada Agricultural University	Parbhani	Maharashtra
24	Odisha Univ of Agriculture and Technology	Bhubaneswar	Odisha
25	Punjab Agricultural University	Ludhiana	Punjab
26	Maharana Pratap Univ of Agriculture and Technology	Udaipur	Rajasthan
27	Rajasthan Agricultural University	Bikaner	Rajasthan
28	Agriculture University	Kota	Rajasthan
29	Sri Karan Narendra Agriculture University	Jobner	Rajasthan
30	Agriculture University	Jodhpur	Rajasthan
31	Tamil Nadu Agricultural University	Coimbatore	Tamil Nadu

Sl.No.	Name of University	Location	State
32	Chandra Shekar Azad University of Agriculture and Technology	Kanpur	Uttar Pradesh
33	Narendra Deva University of Agriculture and Technology	Faizabad	Uttar Pradesh
34	Sardar Ballabh Bhai Patel Univ of Agriculture and Technology	Meerut	Uttar Pradesh
35	Manyavar Kashiram University of Agriculture and Technology	Banda	Uttar Pradesh
36	Govind Ballabh Pant University of Agriculture and Technology	Pantnagar	Uttarakhand
37	Bidhan Chandra Krishi Viswavidyalaya	Kalyani	West Bengal
38	Uttar Banga Krishi Viswavidyalaya	Cooch Bihar	West Bengal

B: Horticultural Universities

Sl.No.	Name of University	Location	State
39	Andhra Pradesh Horticultural University	Tadepalligudem	Andhra Pradesh
40	University of Horticultural Sciences	Bagalkot	Karnataka
41	Dr. Yashwant Singh Parmar University of Horticulture and Forestry	Solan	Himachal Pradesh
42	Dr. YSR Horticultural University	Veketaramanna-gudem	Andhra Pradesh
43	VCSG Uttarakhand University of Horticulture and Forestry	Bharsar	Uttarakhand

Appendix 2
List of ICAR Crop Research Institutes Having Biotechnology Centre

A. Crop Research Institutes

Sl.No	*Name of Research Institute*	*Location*	*State*
1	ICAR-Central Institute for Cotton Research	Nagpur	Maharashtra
2	ICAR- Central Institute for Jute and Allied Fibres	Kolkata	West Bengal
3	ICAR-National Rice Research Institute	Cuttack	Odisha
4	Central Tobacco Research Institute	Rajamundry	Andhra Pradesh
5	ICAR- Directorate of Groundnut Research	Junagarh	Gujarat
6	ICAR-Indian Institute of Oilseeds Research	Hydereabad	Telangana
7	ICAR- Directorate of Rapeseed-Mustard Research	Bharatpur	Rajasthan
8	ICAR-Indian Institute of Rice Research	Hyderabad	Talangana
9	ICAR-Indian Institute of Seed Science	Mau	Uttar Pradesh
10	ICAR-Indian Institute of Millets Research	Hyderabad	Telangana
11	ICAR-Indian Institute of Soybean Research	Indore	Madhya Pradesh
12	ICAR-Indian Agriculture Research Institute	Pusa	New Delhi
13	ICAR-Indian Agriculture Research Institute	Hazaribagh	Jharkhand
14	ICAR-Indian Grassland and Fodder Research Institute	Jhansi	Uttar Pradesh
15	ICAR-Indian Institute of Maize Research	Ludhiana	Punjab
16	ICAR-Indian Institute of Agricultural Biotechnology	Ranchi	Bihar
17	ICAR-Indian Institute of Pulses Research	Kanpur	Uttar Pradesh
18	ICAR-Indian Institute of Sugarcane Research	Lucknow	Uttar Pradesh
19	ICAR-Indian Institute of Wheat and Barley Research	Karnal	Haryana
20	ICAR-National Bureau of Agricultural Insect Resources	Bangaluru	Karnataka
21	ICAR-National Bureau of Agriculturally Important Micro-organism	Mau	Uttar Pradesh
22	ICAR-National Bureau of Plant Genetic Resources	Pusa	New Delhi
23	ICAR-National Centre for Integrated Pest Management	Pusa	New Delhi
24	ICAR- National Institute of Biotic Stress Management	Raipur	Chhatishgarh
25	ICAR-National Centre on Plant Biotechnology	Pusa	New Delhi
26	ICAR-Sugarcane Breeding Institute	Coimbatore	Tamil Nadu
27	ICAR-Vivekanand Parvatiya Krishi Anusandhan Sansthan	Almora	Uttarakhand

B. Horticultural Sciences

Sl.No	*Name of Research Institute*	*Location*	*State*
28	ICAR-Central Institute for Arid Horticulture	Bikaner	Rajasthan
29	ICAR-Central Institute for Subtropical Horticulture	Lucknow	Uttar Pradesh
30	ICAR-Central Institute for Temperate Horticulture	Srinagar	Jammu and Kashmir
31	ICAR- Central Island Agricultural Research Institute	Portblair	Andman and Nicobar
32	ICAR-Central Plantation Crops Research Institute	Kasargod	Kerala
33	ICAR-Central Potato Research Institute	Shimla	Himachal Pradesh
34	ICAR-Central Tuber Crops Research Institute	Trivendrum	Kerala
35	ICAR-Directorate of Cashew Research	Puttur	Karnataka
36	ICAR-Directorate of Floriculture Research	Pusa	New Delhi
37	ICAR-Directorate of Medicinal And Aromatic Plants Research	Anand	Gujarat
38	ICAR-Directorate of Mashroom Research	Solan	Himachal Pradesh
39	ICAR-Directorate of Onion and Garlic Research	Pune	Maharashtra
40	ICAR-Indian Institute of Horticulture Research	Bangalore	Karnataka
41	ICAR-Indian Institute of Oil Palm Research	Pedavegi	Andhra Pradesh
42	ICAR-Indian Institute of Spices Research	Calicut	Kerala
43	ICAR-Indian Institute of Vegetable Research	Varanasi	Uttar Pradesh
44	ICAR-National Research Centre for Banana	Tiruchirapalli	Kerala
45	ICAR-Central Citrus Research Institute	Nagpur	Maharashtra
46	ICAR-National Research Centre for Grapes	Pune	Maharashtra
47	ICAR-National Research Centre for Litchi	Muzaffarpur	Bihar
48	ICAR-National Research Centre for Orchids	Gangtok	Sikkim
49	ICAR-National Research Centre on Pomegranate	Solapur	Maharashtra

*****

Appendix 3
List of International Crop Research Institutes having Biotechnology Centre

Sl.No.	Name of Research Institute/Centre	Founded in	Location	Country
1	International Rice Research Institute [IRRI]	1960	Manila	Philippines
2	International Wheat and Maize Improvement Centre [CIMMYT]	1963	Mexico City	Mexico
3	International Centre for Tropical Agriculture [CIAT]	1967	Cali	Colombia
4	International Institute for tropical Agriculture [IITA]	1967	Ibadan	Nigeria
5	International Potato Centre [CIP]	1971	Lima	Peru
6	International Crop Research Institute for Semiarid Tropics [ICRISAT]	1972	Hyderabad	India
7	Internationa Centre for Agriculture Research in Dry land Areas [ICARDA]	1977	Aleppo	Syria
8	West Africa Rice Development Association [WARDA]	1970	Bouake	Cote divoire
9	International Plant Genetic Resources Institute [IPGRI]	1974	Rome	Italy
10	Centre for International Forestry Research [CIFOR]	1992	Jakarta	Indonesia
11	International Centre for Research in Agroforestry [ICRAF]	1991	Nairobi	Kenya

*****

Key References

A. Molecular Biology

1. Strickberger, M. W. 1985. Genetics. 3rd ed. Macmillan Publishing Co., New York, USA.
2. Watson, J. D. 1976. Molecular Biology of Gene 3rd edition. Benjamin, Menlo Park, California, USA.
3. Friefelder, D. 1987. Molecular Biology 2nd ed. Jones Bartlett Publishers, Inc. USA.
4. Phundan Singh 2015. Molecular Plant Breeding. Kalyani Publishers, New Delhi.
5. Phundan Singh 2018. Plant Breeding: Molecular and New Approaches. Fourth Edition. Kalyani Publishers, New Delhi.
6. Phundan Singh 2012. Molecular Genetics. Kalyani Publishers, New Delhi.
7. Phundan Singh 2012. Molecular Genetics: At A Glance. Kalyani Publishers, New Delhi.
8. Phundan Singh 2012. Objective Molecular Genetics. Kalyani Publishers, New Delhi.

B. Plant Biotechnology

9. Bajaj, Y.P.S. 1994. Somatic Hybridization in Crop Improvement. Springer Verlag.
10. Bhojwani, S. S. (ed.) 1990. Plant Tissue Culture: Applications and Limitations. Elsevier, Amesterdam.
11. Bhojwani, S. S. and Bhatnagar, S. P. 1990. Embryology of Angiosperms. Vikas Publishing Co., New Delhi.
12. Bhojwani, S. S. and Rajdan, M. K. 1996. Plant Tissue Culture: Theory and Practice. Revised Edition. Elsevier, Amesterdam.

13. Chawla, H. S. 2002. Introduction to Plant Biotechnology. 2nd ed. Oxford and IBH Publishing Co.PVT. LTD., New Delhi.
14. Khanna, V. K. 1999. Plant Tissue Culture. Kalyani Publishers, New Delhi.
15. Phundan Singh 2013. Introduction to Biotechnology. Second edition. Kalyani Publishers, New Delhi.
16. Phundan Singh 2019. Plant Biotechnology. Second Edition. Kalyani Publishers, New Delhi.
17. Phundan Singh 2013. Principles of Plant Biotechnology. Kalyani Publishers, New Delhi.
18. Phundan Singh 2016. Objective Plant Biotechnology. Kalyani Publishers, New Delhi.
19. Phundan Singh 2016. Plant Biotechnology At A Glance. Kalyani Publishers, New Delhi.
20. Prakash, J and Pierik, R.L.M. (eds). 1993. Plant Biotechnology: Commercial Prospects and Problems. Oxford and IBH Publishing Co.PVT. LTD., New Delhi.
21. Rajdan, M. K. 1993. An Introduction to Plant Tissue Culture. Oxford and IBH Publishing Co.PVT. LTD., New Delhi.
22. Sensen, C. W. (ed) 2002. Essentials of Genomics and Bioinformatics. Wiley-VCH.
23. Singh, B. D. 2006. Plant Biotechnology. Kalyani Publishers, New Delhi.
24. Vasil, I. K. and Thorpe T. A. 1994. Plant Cell and Tissue Culture. Kluwer Academic Publishers, Dordrecht, Netherlands.
25. Zaitlin, M., Day, P. and Hollaender, A. 1985. Biotechnology in Plant Science: Relevance to Agriculture in the Eightees. Academic Press, N.Y.

Books by the Senior Author

A. Plant Breeding

1. Essentials of Plant Breeding
2. Fundamentals of Plant Breeding
3. Plant Breeding: Molecular and New Approaches
4. Molecular Plant Breeding
5. Principles of Seed Technology
6. Seed Technology: At a Glance
7. Objective Seed Technology
8. Practical in Crop Breeding
9. Practical and Numerical in Plant Breeding
10. Numerical Problems in Plant Breeding and Genetics
11. Objective Science of Plant Breeding
12. Plant Breeding At a Glance
13. Plant Breeding [For Under Graduate Students]
14. Principles of Plant Breeding
15. Cotton Breeding
16. Heterosis Breeding in Cotton
17. Breeding Hybrid Cotton
18. Breeding Transgenic Bt. Cotton
19. Cotton Improvement in India
20. Sankar Kapas [In Hindi]
21. Glimpses of Cotton Breeding
22. Glossary of Plant Breeding and Genetics

23. Elements of Baby Corn
24. Breeding Crop Plants for Stress Resistance
25. Plant Breeding: Related Legislations
26. Concepts in Plant Breeding
27. Commercial Plant Breeding
28. Introduction to Maintenance Plant Breeding

B. Genetics

29. Elements of Genetics
30. Genetics
31. Principles of Genetics [For UG Students]
32. Plant Genetics
33. Fundamentals of Genetics
34. Cotton Genetics
35. Objective Genetics
36. Genetics: At a Glance
37. Objective Genetics and Plant Breeding
38. Objective Genetics and Plant Breeding [Hindi Edition]
39. Glossary cum Dictionary of Genetics and Plant Breeding
40. Molecular Genetics [Objective]
41. Molecular genetics [At a Glance]
42. Molecular Genetics [Subjective]
43. Genetics and Man
44. Genetics Today

C. Quantitative Genetics

45. Biometrical Techniques in Plant Breeding
46. Application of Biometrical Techniques in Plant Breeding [Hindi Edition]
47. Objective Quantitative Genetics
48. Quantitative Genetics: At a Glance
49. Quantitative Genetics

D. Plant Biotechnology

50. Introduction to Biotechnology
51. Plant Biotechnology
52. Principles of Plant Biotechnology
53. Plant Biotechnology: At a Glance
54. Objective Plant Biotechnology
55. Molecular Biology and Plant Biotechnology

E. Intellectual Property Rights

56. IPR and Plant Breeders' Rights [Subjective]
57. IPR and Plant Breeders' Rights [Objective]
58. IPR and Plant Breeders' Rights [At a Glance]
59. Introduction to Intellectual Property Rights
60. Intellectual Property Rights: At a Glance
61. Intellectual Property Rights: Objective

F. Competitive Examination

62. A Guide to Competitive Examinations of Agriculture
63. Objective General Agriculture for Competitive Examinations

*****

Index

www.ingramcontent.com/pod-product-compliance
Ingram Content Group UK Ltd.
Pitfield, Milton Keynes, MK11 3LW, UK
UKHW021947270726
14060UKWH00002B/400